PETRI NETS IN SCIENCE AND ENGINEERING

Edited by **Raul Campos-Rodriguez**
and **Mildreth Alcaraz-Mejia**

Petri Nets in Science and Engineering
http://dx.doi.org/10.5772/intechopen.72258
Edited by Raul Campos-Rodriguez and Mildreth Alcaraz-Mejia

Contributors

José Carlos Quezada Quezada, Ernesto Flores García, Joselito Medina Marín, Jorge Bautista López, Víctor Quezada Aguilar, Wlodek Zuberek, Roman Stryczek, Ivo Martiník, Hamdi Awad, Jianing Wu, Raul Campos-Rodriguez

First published in London, United Kingdom, 2018 by IntechOpen
IntechOpen is the global imprint of INTECHOPEN LIMITED, registered in England and Wales, registration number: 11086078, The Shard, 25th floor, 32 London Bridge Street
London, SE19SG – United Kingdom
Printed in Croatia

British Library Cataloguing-in-Publication Data
A catalogue record for this book is available from the British Library

Additional hard copies can be obtained from orders@intechopen.com

Petri Nets in Science and Engineering, Edited by Raul Campos-Rodriguez and Mildreth Alcaraz-Mejia
p. cm.
Print ISBN 978-1-78923-692-7
Online ISBN 978-1-78923-693-4

For EU product safety concerns:
IN TECH d.o.o., Prolaz Marije Krucifikse Kozulić 3, 51000 Rijeka, Croatia,
info@intechopen.com or visit our website at intechopen.com.

Meet the editors

Raul Campos-Rodriguez is a part-time professor at the Monterrey Institute of Technology and Higher Education in Mexico. He received his BE degree in Computer Systems from the University of Guadalajara (2000) and his master and PhD degrees in Electrical Engineering from the CINVESTAV-IPN (2002 and 2007) in Jalisco, Mexico. From 2007 to 2011, he was a research professor at the University of Guadalajara. From 2011 to 2017, he was a research professor at the ITESO University. He has published a number of international journal papers, an international patent, and directed research and development projects in the area of discrete-event systems, control, observability, embedded systems design and implementation and machine learning algorithms for predictions on dynamic software environments.

Mildreth Alcaraz-Mejia is a research professor at ITESO University. She received her BE degree in Computer Systems from the University of Colima (2001) and her master and PhD degrees in Electrical Engineering from the CINVESTAV-IPN (2003 and 2007) in Jalisco, Mexico. From 2007 to 2011, she was a research professor at the University of Guadalajara. From 2011, she has been a research professor at the ITESO University. She has published numerous scientific articles that have been internationally refereed, book chapters, and one patent in the USA. She has directed research and development projects in the area of discrete-event systems, high performance computing and big data analysis. Her main interest areas include discrete-event systems, design, HPC, data mining and big data analysis, as well as methods for scheduling problems.

Contents

Preface

Petri Nets have been widely used in different areas of science and technology since they were first presented in the middle of the last century by Dr. Carl Adam Petri. It is believed that he planned to use Petri Nets to represent chemical reactions.

Since then, Petri Networks have gained great acceptance in the design of manufacturing processes, logistic processes, software design, communication protocols, modern control theory, artificial intelligence and, of course, chemical process simulations.

Dra. Alcaraz-Mejia and I have been involved with Petri Nets since our master and PhD degrees from 2000 onwards. We used Petri Nets as a way to represent the dynamics of abstract discrete-event systems. Since then, we have used this modeling tool in our research activities in the areas of fault prediction and detection, observability and controllability analysis, to mention a few. Also, we have used Petri Nets as an instrument to teach the student models of concurrency, programming for embedded systems, distributed systems, data modeling and so on.

The graphic nature of the models in Petri Nets makes their use in different areas of science and engineering very attractive, since the models can be obtained in a very intuitive and natural way. Additionally, Petri Nets have a solid mathematical base that comes from linear algebra, vector spaces and linguistic analysis, among others. This allows a formal analysis of important properties such as liveness, interlocking, reachability, controllability or observability, to mention a few.

The aim of this book is to present a collection of works to show the usability of Petri Nets in different areas of science and engineering. Whenever possible, models and graphics illustrate the concepts developed in the chapters. The focus was to develop intuitive ideas and to motivate the reader, as much as possible, to use Petri Nets

We would like to thank Ms. Lada Bozic, Author Service Manager from INTECHOPEN, for her dedication and support during this process. With her help and coaching, we had incredible guidance to do our best effort to assure the quality of this book.

We hope that the readers found this book entitled "Petri Nets in Science and Engineering" illustrative and useful for their work in science and engineering.

Dr. Raul Campos-Rodriguez
Part-time Professor
School of Engineering and Sciences
Western Region
Monterrey Institute of Technology and Higher Education, Jalisco, Mexico

Dra. Mildreth Alcaraz-Mejia
Research Professor
Electronics, Systems and Informatic Department
ITESO University, Jalisco, Mexico

Chapter 1

Introductory Chapter: Petri Nets in Science and Engineering

Raul Campos-Rodriguez and Mildreth Alcaraz-Mejia

Additional information is available at the end of the chapter

http://dx.doi.org/10.5772/intechopen.79309

1. Introduction

The Petri nets are one of the most widely used methods for the study of the dynamics that falls within the category of Discrete Event Systems (DES) [1]. The DES is a class of systems that are guided by the occurrence of events asynchronous in time, which are becoming more and more relevant nowadays. The Petri nets are graphically represented as a directed graph, with two classes of nodes, called places and transitions. The places allow capturing the state of a system. They also represent the conditions required by the events to occur, or to execute, in the DES.

The transitions represent the events, or actions, executed in a system. The execution of the transitions may require one or more conditions to be activated. Moreover, it is possible that a transition does not include input places, as t_1 in **Figure 2**. This class of transitions allows capturing situations in a DES where an event may be random or stochastic, for example, the arrival of an information package in a communication channel. The explanation of the Petri net in **Figure 2** will be addressed lather in this section, after the introduction of the system that it represents.

Figure 1 depicts a conceptual diagram of a multitasking manufacturing system [6]. The system is supplied with the raw material from two conveyors, C_1 and C_2. A robot arm distributes the raw material to either a mill machine or to a lather machine, depending on the manufacturing recipe. The semi-finished pieces are then moved by transporting bands to the assembly machine.

Figure 2 depicts a Petri net model for this multitasking manufacturing system. The supply of the raw material is represented as two transitions with no inputs. In means the material may arrive at any time that the inventory of raw pieces is able to feed the manufacturing system. The robotic arm moves the raw pieces to the mill machine, by means of t_4, or to the lather machine, by means of t_5. The semi-finished pieces are moved to the assembly station to produce a final product.

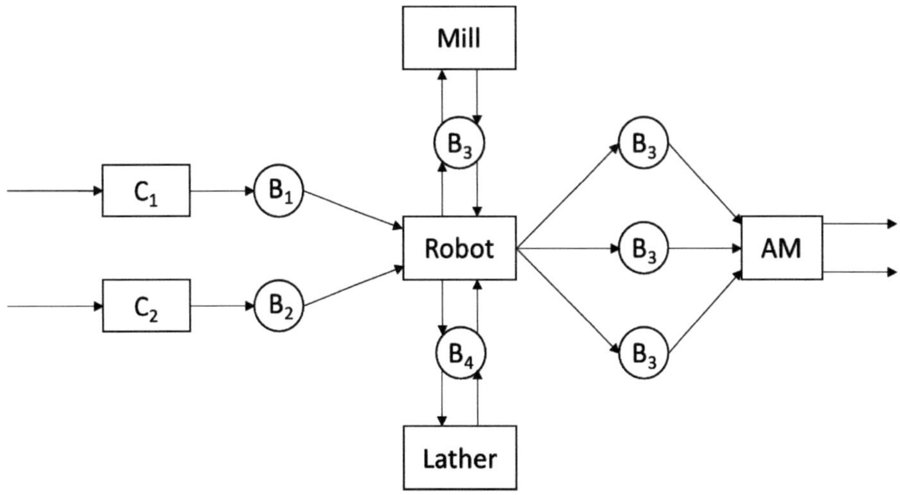

Figure 1. A multitasking manufacturing station. The system is supplied with raw material from two inventories C_1 and C_2. A robotic arm moves the raw pieces to the mill machine or to the lather machine depending on a manufacturing recipe. The robotic arm then moves the semi-finished pieces to the assembly station (AS), where final products are produced.

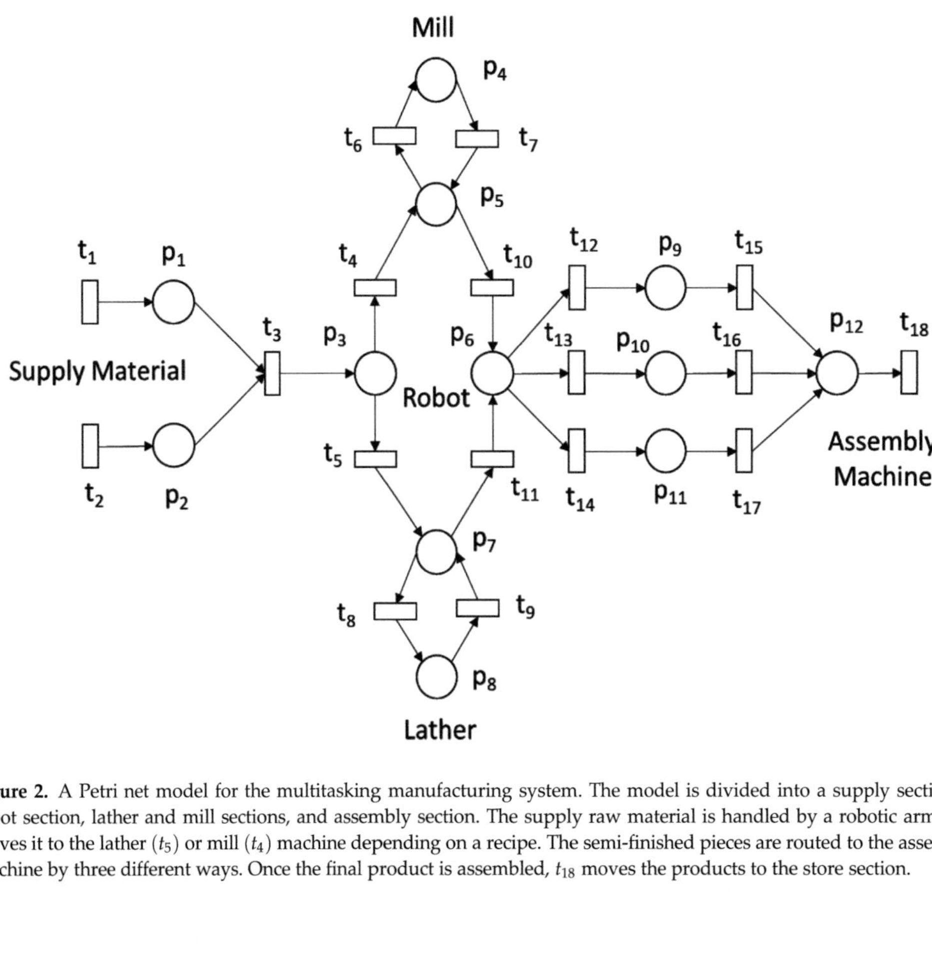

Figure 2. A Petri net model for the multitasking manufacturing system. The model is divided into a supply section, a robot section, lather and mill sections, and assembly section. The supply raw material is handled by a robotic arm that moves it to the lather (t_5) or mill (t_4) machine depending on a recipe. The semi-finished pieces are routed to the assembly machine by three different ways. Once the final product is assembled, t_{18} moves the products to the store section.

Depending on what is the interest of study of the system, for example, the design of control strategies or the evaluation of performance of the assembly recipe, the model in **Figure 2** could be refined or extended. Even new sections of the assembly system may be added to the Petri net model.

The manufacturing system and assembly lines, as well as communication protocols, are some of the most popular type of systems that are modeled and studied with Petri net models [8–10]. However, other types of systems such as workflow management or logistic systems are similarly likely to be modeled and studies by means of Petri nets [15–17]. Moreover, the design and implementation of complex software systems is as well plausible to be addressed with Petri net models [18, 19, 26].

The addressing of software design with Petri nets is popular because the construction of models for complex structures and control flow is quite intuitive thanks to its graphical nature. Moreover, the techniques developed around the Petri nets allow the construction of models that are usually more compact than the produced by other methods, such as those developed in graph theory. However, Petri nets and graph theory are not antagonist. On the contrary, the theory developed in one of them is usually extended to the other. Thus, they are usually complementary to each other.

Figure 3 depicts a block diagram of a reader and writer problem in computer sciences. The processes share a region of memory where they can read and write. The diagram depicts the process that can read, process that can write, and process that perform both operations, reading and writing to the shared memory region. This situation arises in several cases in the design of monolithic and distributed system, within the area of software design.

Figure 4 depicts a model for the above problem of readers and writers [2]. The net represents a system with $2k$ readers modeled as p_2. The system allows up to k parallel reads from a shared memory region. It is represented by the marking in p_3. However, the writing operation

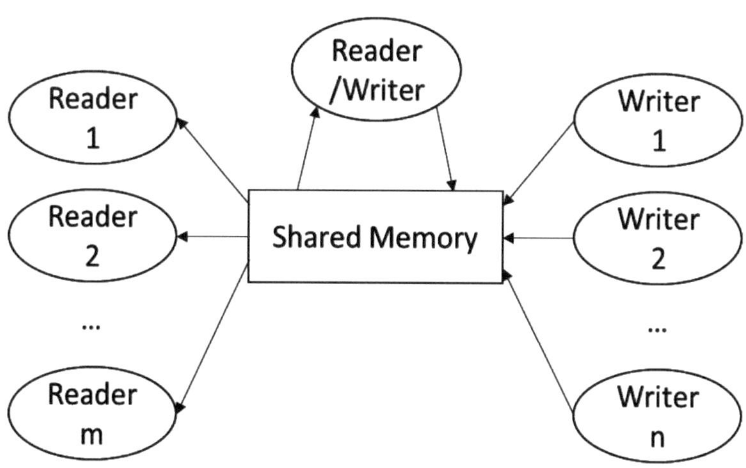

Figure 3. The problem of readers and writers. The problem considers a set of process that can read and write to a shared memory region. The system must allow any number of simultaneous reading operations, while the writing operation requires that no reading operation is in execution. On the other hand, when a writing operation is in execution, no other writing, nor reading, operation is allowed.

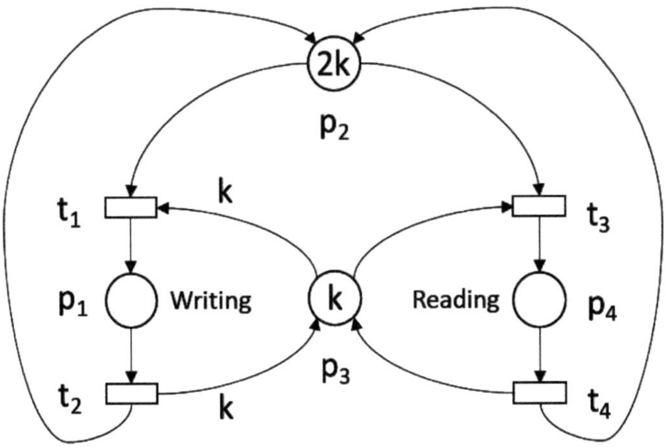

Figure 4. A Petri net model that represents the typical problem of readers and writers. The model allows up to $2k$ processes (p_2) that can read or write to and from shared memory resources (p_3). However, only k of them can be concurrently in a reading operation. On the other hand, the writing process requires k tokens to be on the shared place p_3. That is, no reading operation must be executed by any of the $2k$ readers in p_2 to allow the writing operation. Thus, the writing operation must wait until all the reading operations have finished. Similarly, when a writing operation is in execution, no reading operation is allowed, since p_3 is empty.

requires k tokens on p_3. That is, it requires that no reading operation is currently in execution. Correspondingly, when a writing operation is in execution, no read operation is allowed. This is represented by k weighted arc of t_1. Thus, when writing operation is in execution, by the firing of t_1, the k tokens in p_3 are removed. Once the reading operation is done, the firing of t_2 returns k tokens to p_3. The Petri net model allows any of the $2k$ processes to read and to write to the shared place p_3 by connecting the place p_2 to the reading or writing sections of the net.

Other attractive attribute of the Petri nets is their solid mathematical basis. The incidence matrix that represents the structure of the net in **Figure 4** is represented by Eq. (1). The incidence matrix is independent of the initial condition of the net. This structure could be analyzed by methods from the matrix theory, linear algebra, or vector spaces, for example.

$$\begin{bmatrix} 1 & -1 & 0 & 0 \\ -1 & 1 & -1 & 1 \\ -k & k & -k & k \\ 0 & 0 & 1 & -1 \end{bmatrix} \tag{1}$$

The state equation of a Petri net allows a formal definition of its dynamics. The next state of a Petri net can be computed from the current state, and a multiplication of the matrix that represents the structure of the net and a vector that represents the transitions that can fire, as follows:

$$M_{k+1} = M_k + B\vec{u_k} \tag{2}$$

The vector $\vec{u_k}$ represents one or more transitions that are allowed to fire. It is known as the Parikh vector, in a clear relationship to the Parikh's theorem. This theorem relates the strings in a context-free language and the number of the occurrences of the symbols in these strings.

In a similar way, the vector $\vec{u_k}$ represents the number of times each transition is fired at a given stage in the evolution of the net. In this sense, the Parikh vector behaves like a "functor" in the sense of the category theory [3], from the strings over the alphabet of events in a DES to vectors that quantifies the occurrence of events in a DES. That is, the Parikh vector "loses" the execution order of the events in a trajectory of a DES to obtain a pure vector which is simpler to operate by a matrix multiplication.

There are different semantics for the execution of the transitions in a Petri net model. First, in a single firing semantics, only one of the enabled transitions can fire at a time. Second, in the multiple firing semantics, all the enabled transitions are allowed to fire at a time. In all the semantic approaches, the conflicting transitions, that is, the ones whose firing disables the firing of others are resolved by priorities, by a probability distribution, or some other conflict resolution mechanism. Depending on the adopted semantics, the ability of the models to capture dynamics of real systems differs. For example, if the analyzed system is of distributed nature, such as a cluster of computers or a cloud service, then the correct semantics is that of multiple firing. The expressiveness of the different semantic mechanisms is a theoretical question that lies around the computer sciences. The next subsections detail some theoretical aspects of the Petri nets and application in science and theory of systems.

2. Petri nets in science

As mentioned, the Petri nets are a very versatile tool that turns it useful in science, as well as in engineering. In the science field, a wide developed aspect is related to the study of Petri nets as a system and their associated abstract properties.

For example, the study of properties of the Petri net models in terms of vectors and matrices is complemented with the linguistic study in terms of strings and formal languages. The well-developed theory of matrices, linear algebra, and vector spaces are well suited to the analysis of properties in the net models, providing efficient solutions. However, other studies such as the reachability analysis, requires the partial expansion of the state space of the models, which turns the investigation in an inverse direction, from a vector space to string over a language. Though, the advantages of the study of the Petri net properties in terms of vectors, matrices, and linear algebra in general are considerable, and many of the theory developed for Petri nets relies on them.

Indeed, by restricting the marking of a Petri net to be non-negative, the state space entirely lies in the positive cone of Z^+. Thus, some of the theory of positive linear systems could be applied [7]. **Figure 5(a)** depicts the state space, in R^3, of the Petri net model in (b) for two different initial conditions $M_0 = [1\ 0\ 0]$ and $M_0 = [2\quad 0\quad 0]$, the two hyperplanes are orthogonal to the unitary vector $u = [1\quad 1\quad 1]$. Moreover, if the net is conservative (i.e., the number of tokens over all its places remains constant for any evolution trajectory), then it is easy to show that the entire state space of the Petri net lives in one of the hyperplanes orthogonal to a vector in R^n, where n is the number of places of the net.

One of the most active areas of the applications of the Petri nets in science is in the field of the modern control theory. The study of control techniques for discrete event system, including Petri net models, covers the range of applications from design of discrete event controller, design of state observers, analysis of fault tolerant systems, analysis of Lyapunov-like stability, detectability analysis, isolation, and failure recovery techniques, among others [4–6].

In this context, considering a Petri net as an input-output system, as in classical control theory, is useful. **Figure 6** depicts the state equation of a Petri net given by a block diagram. The input vector $\vec{u_k}$ is operated by the matrix B to produce a marking increment that is added to the current marking. The sum of these two quantities becomes the new marking reached by the current evolution of the net. A unit delay block allows the new marking becoming the current marking for the next evolution of the net.

Considering the state equation of a Petri net as a block diagram as in **Figure 6** allows studying the dynamics of the model as in the control theory [4]. Techniques for the construction of feedback controller or state observers could be addressed [6]. Performance analysis is as well, a usual analysis stage in the design of the class of systems that could be modeled by a Petri net [5].

Figure 7 depicts a block diagram of a classical control scheme for a DES modeled as a Petri net. The scheme considers two models, one of the system and the other of a reference. The controller receives the difference of the output of the system and the reference in order to compute the control actions. The objective of the control scheme is to achieve zero error, by the actions that the controller can exert over the system. If the system to be controlled is a software,

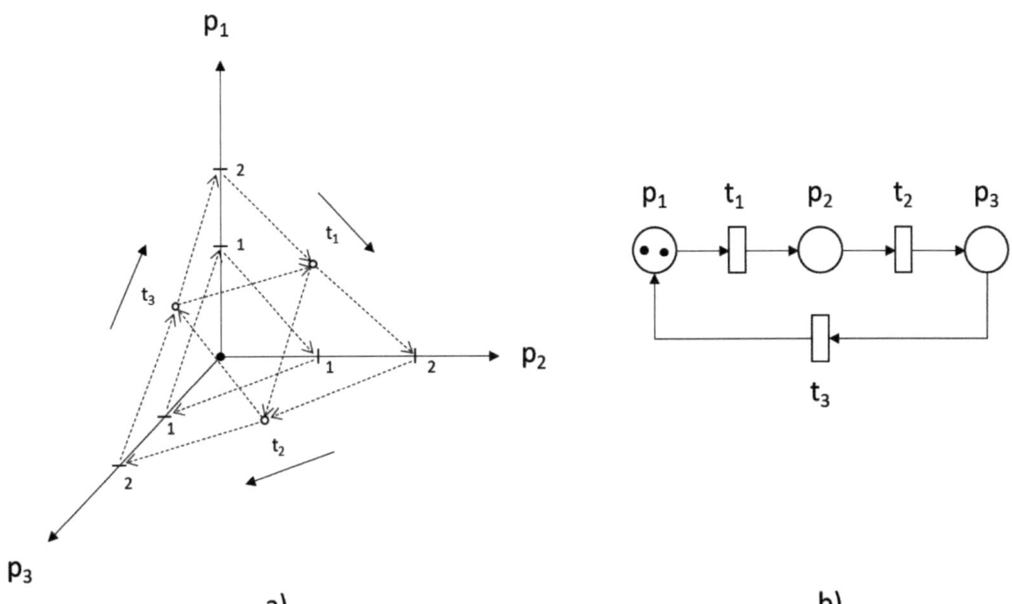

Figure 5. A Petri net model and its respective state space shown as a hyperplane. The solid arrows show the flow of the marking by the firing of the corresponding transition. Dashed arrows represent the marking change by the firing of a single transition. Increasing the number of tokens in the initial marking represents an orthogonal movement of the hyperplane away of the origin. The figure illustrates two hyperplanes. The lower one represents the initial marking with one token at p_1, while the upper one represents the initial marking with two tokens as p_1.

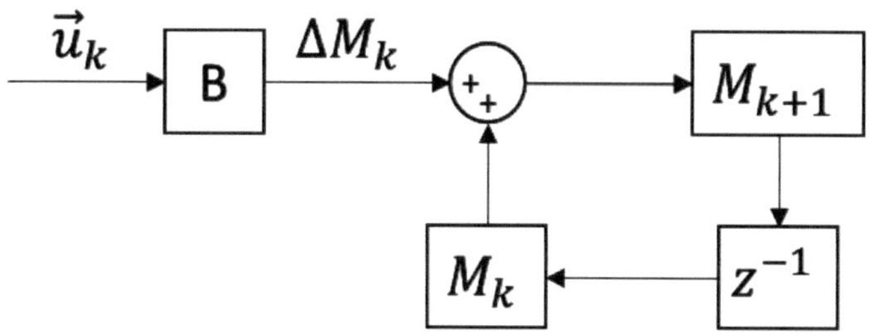

Figure 6. The state equation of a Petri nets as a block diagram. The input vector u_k is multiplied by the matrix B to produce an increment of marking ΔM_k. This increment is added to the current marking M_k to produce the new marking M_{k+1}. This new marking becomes the current marking for the next evolution of the net.

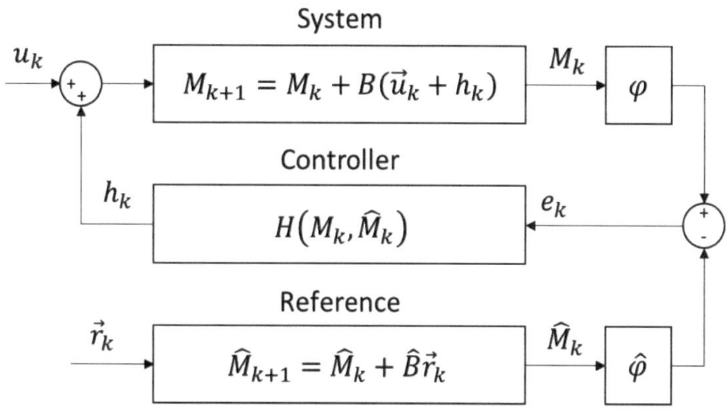

Figure 7. A control scheme for Petri nets. The system is modeled as a Petri net model. The required behavior for the system is modeled as other net model called reference. The objective of the controller is to achieve zero error in the difference of the outputs of the system and reference. If the system is a database software, for example, then the reference is a specification that the database manager requires in the database. The controller is another software that makes the overall scheme autonomous.

for example, either distributed or monolithic, then the reference is a specification or recipe that the software must meet. Then, the controller is another software responsible for computing the required parameters and configurations in order to adapt the main software system to the required behavior in an autonomous fashion. There is a huge trend in cloud computing and artificial intelligence to transform current software systems, such as database clusters, into autonomous intelligent systems that automatically adapts to user requirements and that are even able to predict future workloads and adapt to them [11, 12].

The next section reviews some illustrative use of Petri net models in engineering applications.

3. Petri nets in engineering

The usability of the Petri nets in engineering applications is as well widely accepted. The stages of the design, the implementation, and the validation of systems are suitable addressed with

Petri net models. The covered applications include communication protocols, distributed systems, distributed database, concurrent and parallel programming and systems, industrial control systems, multicore processor platforms, dataflow-computer systems design, workflows and process-driven systems, fault-tolerant systems, and to mention a few. Properties practical interest such as fairness in the execution of tasks, deadlock avoidance, state reachability, process interlocking, among others, are possible to be analyzed within the Petri net framework.

For example, **Figure 8** illustrates a very simple and conceptual communication protocol. The communication act is analyzed from the sending process point of view. A sender process sends a message by the output buffer and blocks its activity while waiting for an acknowledgement by the input buffer. A receiver process is blocked while waiting for an input message. Once a message has arrived, the receiving process reads the message and sends an acknowledgment by the input buffer. After the communication act has finished, the process restarts its logic to be ready for the next communication. This model could be extended to include faulty communication channels, which may lose the messages, acknowledge expiration periods, or other characteristics of practical interest. The analysis and design of communications protocols has been widely addressed with the use of Petri net models [9, 10].

There are some extensions to the Petri nets to handle specific aspects of different engineering problems. Some of the extensions add structure and information to the tokens, transitions, and places of a net. These extensions allow the construction of models that are quite compact

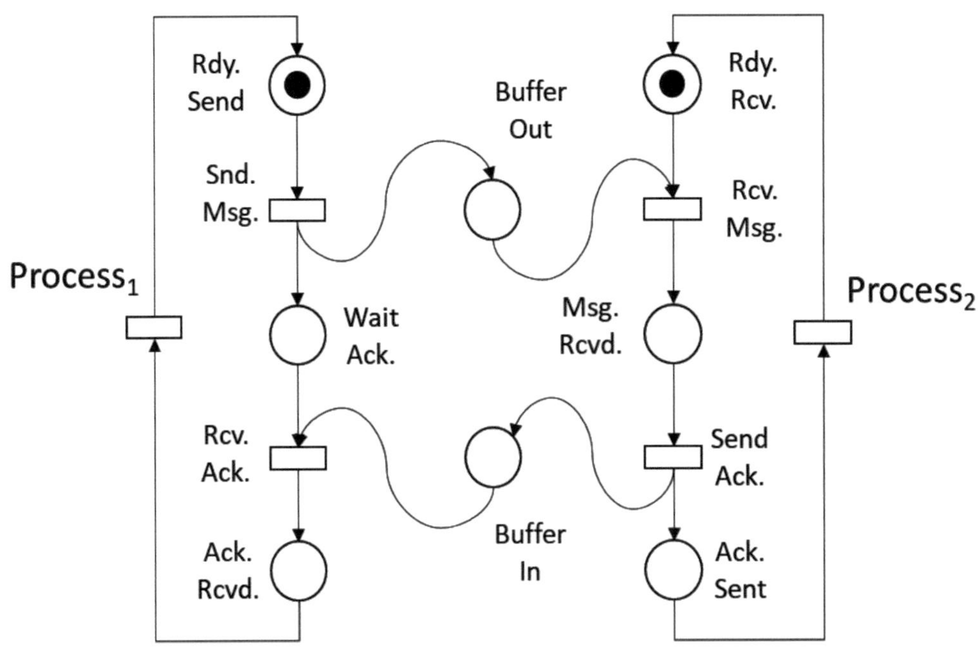

Figure 8. A Petri net representing a communication protocol. Process 1 sends a message by the output buffer and waits for an acknowledgement buy the input buffer. Process 2 reads the message from the output buffer and sends an acknowledgment by the input buffer. After the communication protocol is completed, the both processes restart their logics to get ready for the next communication act.

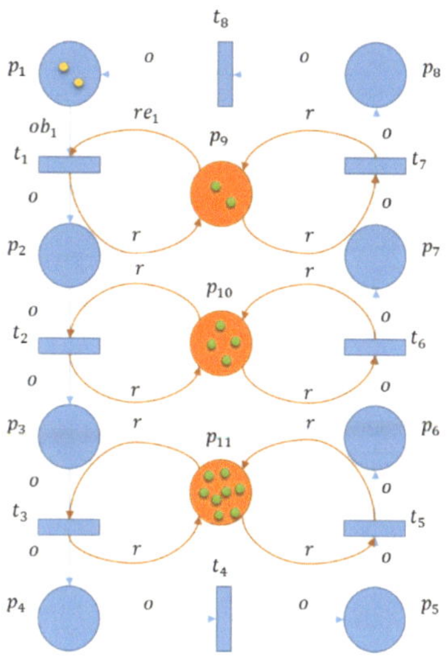

Pre	t_1: f_1	t_2: f_1	t_3 f_1	t_4 f_2	t_5 f_3	t_6 f_3	t_7 f_3	t_8 f_4
$p_1:o$	si_1							
$p_2:o$		si_1						
$p_3:o$			si_1					
$p_4:o$				si_2				
$p_5:o$					si_3			
$p_6:o$						si_3		
$p_7:o$							si_3	
$p_8:o$								si_4
$p_9:r$	ri_1						ri_2	
$p_{10}:r$		ri_1				ri_2		
$p_{11}:r$			ri_1		ri_2			

Post	t_1: f_1	t_2: f_1	t_3 f_1	t_4 f_2	t_5 f_3	t_6 f_3	t_7 f_3	t_8 f_4
$p_1:o$								so_4
$p_2:o$	so_1							
$p_3:o$		so_1						
$p_4:o$			so_1					
$p_5:o$				so_2				
$p_6:o$					so_3			
$p_7:o$						so_3		
$p_8:o$							so_3	
$p_9:r$	ro_1						ro_2	
$p_{10}:r$		ro_1				ro_2		
$p_{11}:r$			ro_1		ro_2			

Figure 9. A colored Petri net model of a task scheduling problem. The structure of the net represents the different stages of the processes in a multitasking environment. The working processes compete among them to acquire the jobs that need to be executed. Each token is a composite unit that carries the information about the state of the working process. The simulation of this model allows us to study the performance of different scheduling policies.

compared to the models obtained with the traditional approach. These models are called Colored Petri Net (CPN) [18].

As an illustration, **Figure 9** depicts a CPN for a task scheduling problem. The structure of the net represents the different stages of the working processes in a distributed multitasking environment. The left-hand side of the net structure represents the stages that the processes perform to acquire a job. The right-hand side represents the stages that the working processes perform to release the resources and update the state of the overall scheduling problem. The simulation of the model depicted in the figure allows to study the performance of different scheduling policies over different workload conditions. For example, it is possible to approximate the optimal number of process required by the scheduling problem for a fixed number of tasks. Even more, it is possible to study an optimum rate in the increment of the working processes given a rate in the increment of the tasks over discrete interval of times [27].

Recently, with the increase of the cloud computing and the massive data content in the social networks, the machine learning techniques and the methods related to the data analytics are essential tools in the study and investigation of the big data. There are several proposals to allow the Petri net models learn some kind of fuzzy reasoning and decision making [22–24]. Similar approaches as that of the supervised and unsupervised learning have been addressed [21, 25].

Figure 10 shows a Petri net representing a workflow pattern for a customer reclaim system. The customer may initiate a request at any time. Two activities are launched in parallel once a request is in the system. First, a ticket check process is executed. Second, an examination of the request is performed. At this point, based on the machine learning, data analytics and/or

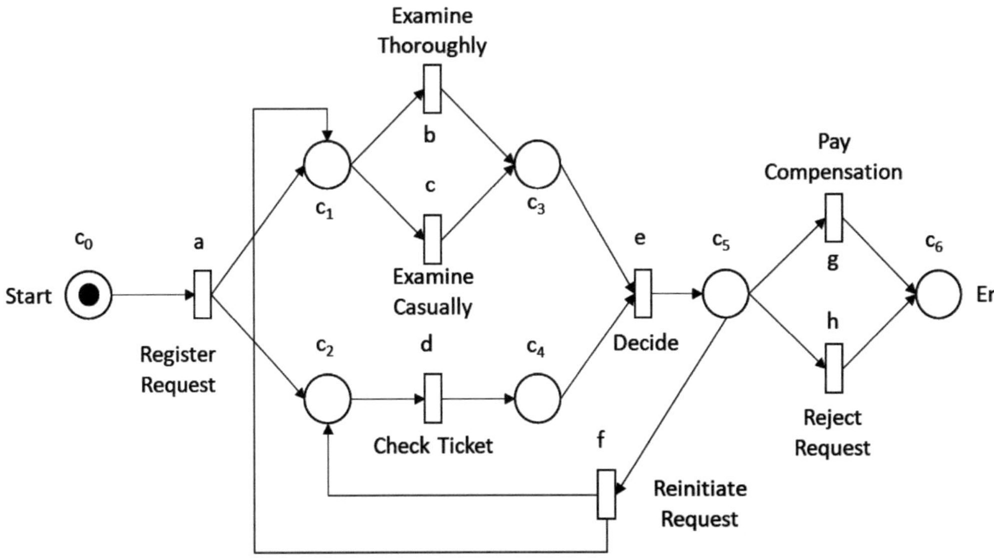

Figure 10. A Petri net model for a workflow pattern of a customer request. The customer may initiate a new request at any time. Two activities are launched in parallel for every request. One activity is to check the ticket. The other is to examine the request. At this point, a decision is made about how deep to examine the request. At this point, a guard based on the machine learning and data analytics is constructed for the transitions b and c. A decision process then comes, where either a compensation payment, a request rejection, or a request reinitiating may apply.

statistical learning mechanisms, guards for the transitions b and c are constructed. The guards allow deciding when it is more convenient to execute an in-deep examination process or a casual examination process. On the one hand, it saves time by executing a casual examination when the guard determines that it is more likely that the characteristics of the request are that of a genuine customer request. On the other hand, it saves money by executing a thoroughly examination process when the guard determines that the current request is more likely to be a fraudulent request.

Other important area of the engineering where the Petri nets have been successfully used is in the automatic code generation. The exponential growth of the cloud computing and proliferation of solutions based on the Internet of Things have made the design of the system software supporting them become more challenging. The set of requirements that this type of systems must address includes the sensing of signals in soft and hard real-time and the traditional support for media-reach services. This mixture of requirements turns the design of a correct and efficient system of this type a whole challenge. Approaches based on model-based design promise useful solutions for these challenges. The complex behavior and set of conditions that this class of software must address can be well represented with Petri net patterns. Synchronization mechanisms, message passing, rise conditions, critical sections, parallel and concurrent process, task activation conditions, and user interactions, to mention a few, could be easily represented with intuitive Petri net blocks. Then, a simulation process may allow the study of the performance of the solution and the adjustment of parameters for a fine-tuning process.

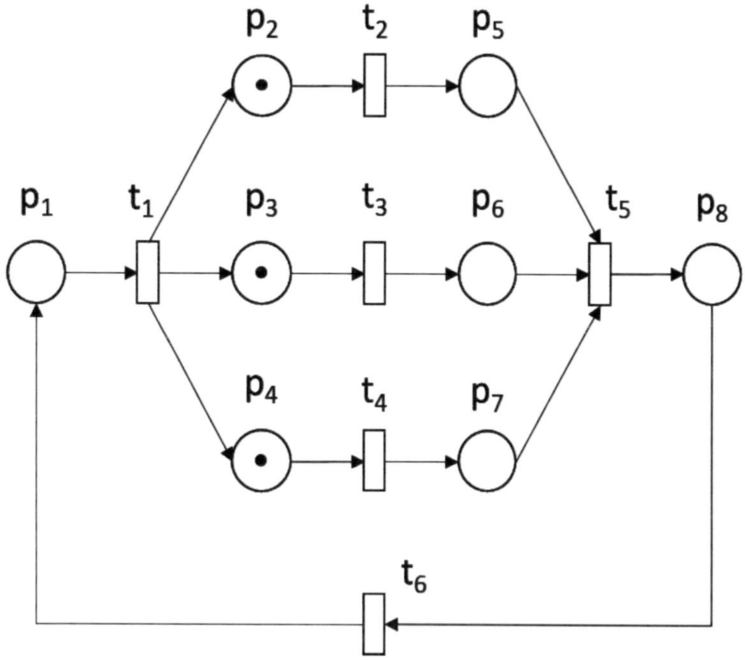

Figure 11. A Petri net model representing three parallel threads. The threads are launched in parallel by the firing of t_1. Each thread runs free to complete its activity. In this design, the end of the threads is synchronized at t_5. That is, if one thread finishes its work before the others, it must wait until the other threads end its activities. Once all the threads have finished, they are reinitialized to repeat the loop.

Figure 11 represents a Petri net model for three parallel processes. The transition t_1 launches the execution of the processes in parallel. Each process runs freely until they end its activities. In this design, the transition t_5 synchronizes the end of the processes. That is, if one process ends its activities, then it must wait the others to end. Once all the processes have finished, the loop repeats infinitely. The transitions t_2, t_3, and t_4 represent the activity load of each process. There are different approaches to add an amount of time to these transitions [13–15]. Within a suitable simulation process, this allows to investigate the performance of the system under different work load conditions, which is a must in the development of real world solutions. Once the parameters of the model have been tuned and its performance evaluated, the next step consists on the synthesis of the code in a target programming language for a specific platform.

For example, **Figure 12** shows a section of code in C/C++ implemented from the model in **Figure 11**. The code implements a set of joinable posix threads. A for loop launches a number of threads defined by the global constant NUM_THREADS. Other for loop waits for the end of the threads. Once all the threads have finished, the loop repeats indefinitely. The automatic code generation from Petri net models has recently been investigated with promissory results [19, 20].

```
...
while(1){
    // Initialize and set threads as joinable
    pthread_attr_init(&attr);
    pthread_attr_setdetachstate(&attr, PTHREAD_CREATE_JOINABLE);
    for( i = 0; i < NUM_THREADS; i++ ) {
      cout << "main() : creating thread, " << i << endl;
      rc = pthread_create(&threads[i], NULL, wait, (void *)i );
      if (rc) {
         cout << "Error:unable to create thread," << rc << endl;
         exit(-1);
      }
    }
    // free attribute and wait for the other threads
    pthread_attr_destroy(&attr);
    for( i = 0; i < NUM_THREADS; i++ ) {
      rc = pthread_join(threads[i], &status);
      if (rc) {
         cout << "Error:unable to join," << rc << endl;
         exit(-1);
      }
      cout << "Main: completed thread id :" << i ;
    }
    cout << "Re-staring the main infinite loop" endl;
}
...
```

Figure 12. A section of code in C/C++ implemented from the Petri net in the figure above. An infinite loop initializes a set of joinable threads. A for loop launches the number of threads specified by the constant NUM_THREADS. A for loop waits for the end of all the launched threads. This cycle repeats forever.

4. Conclusions

This chapter aims to briefly review the applications of the Petri nets in science and engineering. It not pretends to be a deep review of the applications with complete detail and mathematical foundation. Rather, the objective is to provide an illustrative introduction to Petri nets and its potential applications, as intuitive as possible, avoiding the use of complex mathematical notation and formulation. The focus of this chapter was in the graphical nature of the Petri nets and the intuition about them, and with some emphasis in its mathematical foundation. Also, the intention is that this chapter serves as an introduction to this book entitled Petri Nets in Science and Engineering. The authors hope you find this book illustrative for your different activities in science and engineering.

Sincerely,

R. Campos-Rodriguez, M. Alcaraz-Mejia.

Acknowledgements

The authors want to thank Jose Valerio, from Oracle Guadalajara Development Center, for his valuable comments in the review of this chapter and for his experience and comments in Machine Learning and Data Analytics.

Author details

Raul Campos-Rodriguez[1]* and Mildreth Alcaraz-Mejia[2]

*Address all correspondence to: rr_campos@hotmail.com

1 Monterrey Institute of Technology and Higher Education, Guadalajara Campus, Tlaquepaque, Jalisco, Mexico

2 ITESO University, Tlaquepaque, Jalisco, Mexico

References

[1] Reisig W. Petri Nets: An Introduction. Vol. 4. Luxemburgo: Springer Science & Business Media; 2012

[2] Murata T. Petri nets: Properties, analysis and applications. Proceedings of the IEEE. 1989; **77**(4):541-580

[3] Mac Lane S. Categories for the Working Mathematician. Vol. 5. Berlin: Springer Science & Business Media; 2013

[4] Roxin EO. Control Theory and its Applications. Gordon and Breach; 1997

[5] Cassandras CG. Discrete Event Systems: Modeling and Performance Analysis. Aksen Associates Series in Electrical and Computer Engineering, IFAC Proceedings Volumes. 2000:**33**(13):313-318

[6] Cassandras CG, Lafortune S. Introduction to Discrete Event Systems. Berlin: Springer Science & Business Media; 2009

[7] Farina L, Rinaldi S. Positive Linear Systems: Theory and Applications. New York: John Wiley & Sons; 2011

[8] Viswanadham N, Narahari Y. Performance Modeling of Automated Manufacturing Systems. Englewood Cliffs, NJ: Prentice Hall; 1992. pp. 497-508

[9] Merlin P, Farber D. Recoverability of communication protocols—Implications of a theoretical study. IEEE Transactions on Communications. 1976;**24**(9):1036-1043

[10] Bochmann G, Sunshine C. Formal methods in communication protocol design. IEEE Transactions on Communications. 1980;**28**(4):624-631

[11] Pavlo A, Angulo G, Arulraj J, Lin H, Lin J, Ma L, et al. Self-Driving Database Management Systems. In: CIDR. 2017

[12] Available from: https://www.oracle.com/corporate/pressrelease/oow17-oracle-autonomous-database-100217.html

[13] Wang J. Time Petri nets. In: Timed Petri Nets. Boston, MA: Springer; 1998. pp. 63-123

[14] Popova-Zeugmann L. Time Petri nets. In: Time and Petri Nets. Berlin, Heidelberg: Springer; 2013. pp. 31-137

[15] Ling S, Schmidt H. Time Petri nets for workflow modelling and analysis. In: 2000 IEEE International Conference on Systems, Man, and Cybernetics, Vol. 4. IEEE; 2000. pp. 3039-3044

[16] Yu J, Buyya R. A taxonomy of workflow management systems for grid computing. Journal of Grid Computing. 2005;**3**(3–4):171-200

[17] Van der Aalst WM. The application of Petri nets to workflow management. Journal of Circuits, Systems, and Computers. 1998;**8**(1):21-66

[18] Jensen K. Coloured Petri Nets: Basic Concepts, Analysis Methods and Practical Use. Vol. 1. Berlin: Springer Science & Business Media; 2013

[19] Philippi S. Automatic code generation from high-level Petri-nets for model driven systems engineering. Journal of Systems and Software. 2006;**79**(10):1444-1455

[20] Mortensen KH. Automatic code generation from coloured Petri nets for an access control system. In: Second Workshop on Practical Use of Coloured Petri Nets and Design/CPN; Aarhus, Denmark. October 1999. pp. 41-58

[21] Shen VR, Chang YS, Juang TTY. Supervised and unsupervised learning by using Petri nets. IEEE Transactions on Systems, Man, and Cybernetics-Part A: Systems and Humans. 2010;**40**(2):363-375

[22] Bugarin AJ, Barro S. Fuzzy reasoning supported by Petri nets. IEEE Transactions on Fuzzy Systems. 1994;**2**(2):135-150

[23] Konar A. Machine learning using fuzzy Petri nets. Computational Intelligence: Principles, Techniques and Applications. Berlin Heidelberg: Springer-Verlag, 2005. pp. 521-546

[24] Looney CG. Fuzzy Petri nets for rule-based decision making. IEEE Transactions on Systems, Man, and Cybernetics. 1988;**18**(1):178-183

[25] Bulitko V, Wilkins DC. Machine learning for time interval Petri nets. In: Australasian Joint Conference on Artificial Intelligence. Berlin, Heidelberg: Springer; December 2005. pp. 959-965

[26] Badouel E, Bernardinello L, Darondeau P. Petri Net Synthesis. Heidelberg: Springer; 2015. p. 339

[27] Alcaraz-Mejia M, Campos-Rodriguez R, Caballero-Gutierrez M. Modeling and simulation of task allocation with colored Petri nets. In: Computer Simulation. London: InTech; 2017

Chapter 2

Ladder Diagram Petri Nets: Discrete Event Systems

José Carlos Quezada Quezada,
Ernesto Flores García, Joselito Medina Marín,
Jorge Bautista López and Víctor Quezada Aguilar

Additional information is available at the end of the chapter

http://dx.doi.org/10.5772/intechopen.75753

Abstract

Ladder diagram language (LD) is a common programming language in industry to develop control algorithms of discrete event systems (DESs). Besides, it is one of the five programming languages supported by the International Electrotechnical Commission through the IEC-61131-3 standard. On the other hand, Petri net (PN) theory is both a graphical and mathematical tool used to model discrete event systems, particularly in this study, control lines used in industrial algorithms. Control algorithms in LD are generally developed based on the experience of control system programmers. Therefore, it is still a relevant problem how to formalize the current and new control algorithms. In this chapter, are analyzed lines in LD used more frequently in control algorithms. Additionally, an element-to-element transformation methodology from a LD program to a PN model is proposed.

Keywords: control algorithms, discrete event systems, ladder diagram language, model, petri nets

1. Introduction

LD language is one of the five languages contemplated in the standard IEC-61131-3 [1], its use in the industry is due to its similarity with the electrical diagrams, and its behavior is based mainly on the electromechanical relay, but LD language also has the capacity to include logical functions blocks. The others languages are: function block diagram (FBD), instructions list (IL), structured text (ST) and sequential function char (SFC).

There are two types of control lines that are analyzed and converted into PN structures: the logical AND, OR, AND-OR, auto-loop and interlocking, which have both discrete inputs and outputs. The logic blocks such as timer, counter and comparator have all analog inputs, but their control output is discrete.

The main motive or need to model the control algorithms in LD is because they are developed mainly based on the experience of programmers in industrial control [2, 3], so it is important to propose approaches that help guarantee the safe control algorithms applied in machines or industrial processes, and the theory of PN [4] allows modeling the basic control lines used in the LD algorithms. Different approaches have been presented to provide a solution to analyze, model and simulate control algorithms developed in LD with PN or vice versa [5–11].

Physical or discrete memory signals can have two states (activated or deactivated, 0 or 1, etc.), so, we propose a distribution of these signals to PN structure that can model both states, but only one active at a time. On the other hand, the cyclic operation of PLC generates cyclic evaluation of the control algorithm in function of the states of physical input and memory signals. This behavior must be considered to avoid accumulation of tokens in places of PN structures, for which reason marking conditions are proposed in places that represent physical or memory outputs of PN structures of control lines in LD. Likewise, cyclic evaluation of control lines generates the energized and de-energized behavior of coils; therefore, it is also necessary to restore conditions of PN structures of each control line in LD, conditioning the marking in function of the input places [12, 13].

To convert control lines with analog inputs, places where their marking is a data (color in colored Petri nets) are included [9], which may be changing depending on the logic control algorithm. Conditioned transitions are proposed for their firing depending on the behavior of the control block in respective LD.

Based on analysis of the control lines, we propose the definition of a PN for discrete event systems in LD (LDPN), with which PN structures of control lines in LD are generated.

2. Control lines in LD to discrete event systems

The LD language has as its operating principle the behavior of an electromechanical relay, with the option of including function blocks. The standards IEC-61131-3 define LD like "*modeling networks of simultaneous functioning electromechanical elements, such as relay contacts and coils, timers, counters, etc.*" The control lines analyzed are the logic AND, OR, AND–OR, auto-loop, interlocking, timers, counters and mathematic comparisons. The first five logical have discrete inputs and output. Meanwhile in the logical of timers, counters and mathematical comparisons have analog inputs and discrete outputs.

The run of control algorithm in PLC is cyclic, and it mainly performs five actions such as reading of physical inputs, copy status of physical inputs, evaluation of the control algorithm with previous copy, copy of the status of physical outputs and sending of these statuses to physical modules.

http://dx.doi.org/10.5772/intechopen.75753

2.1. Control lines both discrete inputs and outputs

Figure 1 shows the control line of logic AND, when all contacts In_1, In_2, …, In_n allow electric power flow, then Out1 coil is energized. Eq. (1) is the model corresponding.

$$\text{Out1} = \text{In\_1} \,\&\&\, \text{In\_2} \,\&\&\ldots \text{In\_n} \tag{1}$$

Figure 2 shows the control line of logic OR, when any contact In_1, In_2, …, In_n allows electric power flow, then Out1 coil is energized, its model is stand for the Eq. (2).

$$\text{Out1} = \text{In\_1} \parallel \text{In\_2} \parallel \ldots \text{In\_n} \tag{2}$$

Figure 3 shows the control line of logic AND–OR. When the contacts In_1, In_2, …, In_n or the contacts In_1, In_3, …, In_n allow electric power flow, then Out1 coil is energized. Eq. (3) is the model corresponding.

$$\text{Out1} = (\text{In\_1} \,\&\&\, In\_2 \,\&\&\ldots In\_n) \parallel (In\_1 \,\&\&\, In\_3 \,\&\&\ldots In\_n) \tag{3}$$

Figure 4 shows the control line of logic auto-loop. When the contacts In_1, In_2, …, In_n or the contacts Out1, In_2, …, In_n allow electric power flow, then Out1 coil is energized. Eq. (4) is the model corresponding.

Figure 1. Control line of logic AND.

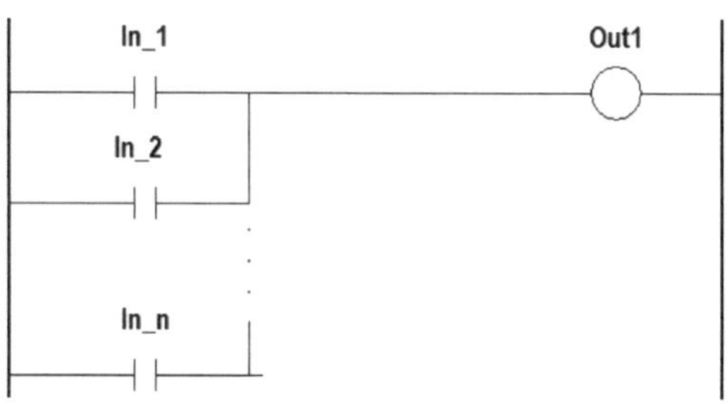

Figure 2. Control line of logic OR.

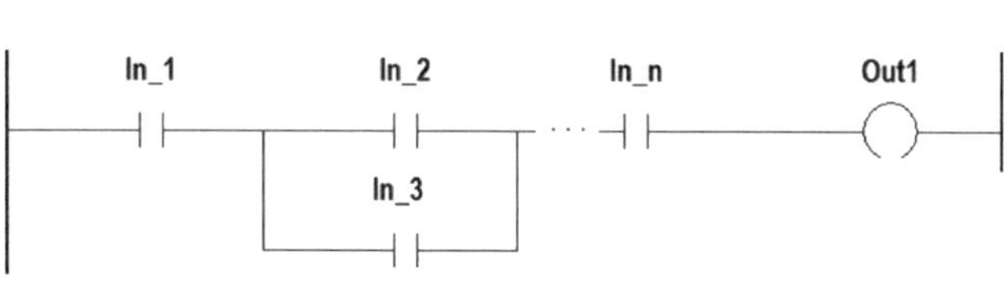

Figure 3. Control line of logic AND–OR.

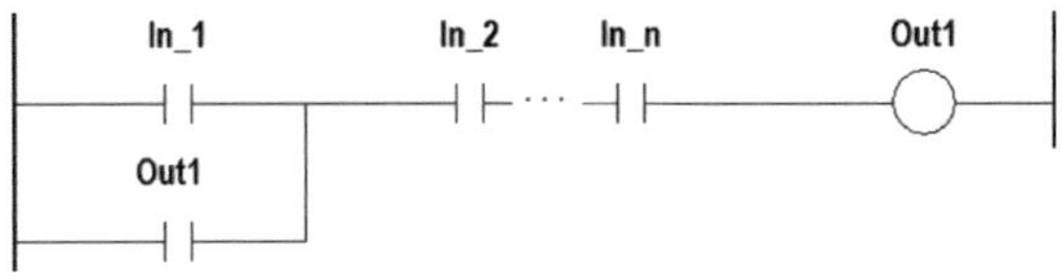

Figure 4. Control line of logic auto-loop.

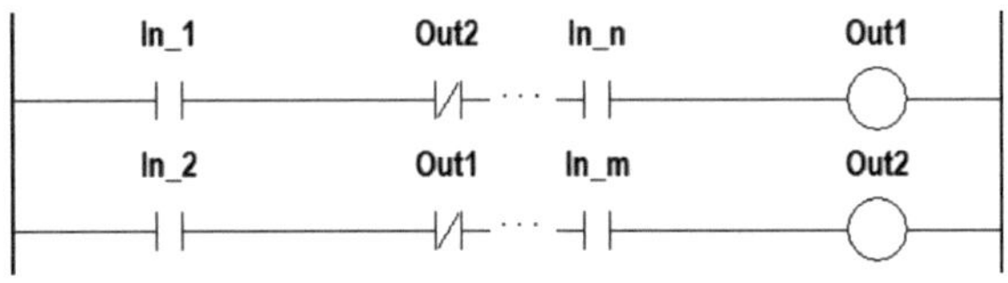

Figure 5. Control line of logic interlocking.

$$\text{Out1} = (\text{In\_1} \,\&\, \&In\_2 \,\&\, \&In\_n) \,||\, (Out1 \,\&\, \&In\_2 \,\&\, \&In\_n) \tag{4}$$

Figure 5 shows the control line of logic interlocking, when the contacts In_1, ~Out2, …, In_n allow electric power flow, then Out1 coil is energized, and it blocks the energizing of Out2 coil. If Out2 coil is energized first, then Out1 coil cannot be energized. Eq. (5) is the model corresponding.

$$\text{Out1} = \text{In\_1} \,\&\, \&\overline{\text{Out2}} \,\&\, \&\ldots\ In\_n; \text{Out2} = \text{In\_2} \,\&\, \&\overline{\text{Out1}} \,\&\, \&\ldots\ In\_m \tag{5}$$

2.2. Control lines with analog inputs and discrete output

Figure 6 shows the standard function block of on-delay timer (TON) and its timing diagram of the functional [1]. The signals *Preset_time* and *Elapsed_time* are analog. If the contact In_1 allows electric energy flow, when *Elapsed_time* adds base time and if *Elapsed_time* is equal or greater than *Preset_time*, then Out1 coil is energized. Eq. (6) depicts the logic model of the block TON.

$$\text{If } (\text{In\_1} = 1 \,\&\, \&\text{ET} \geq \text{PT}), \text{ then Out1} = 1 \tag{6}$$

Restart condition: If In_1 = 0, then ET = 0 and Out1 = 0.

Figure 7 shows the standard function block of off-delay timer (TOF) and its timing diagram of the functional [1]. If the contact In_1 allows energy power, then the Out1 coil is energized, and the *Elapsed_time* variable is set to zero. When the In_1 signal is equal to zero, the *Elapsed_time* variable adds base time and if *Elapsed_time* is equal or greater than *Present_time*, then Out1 coil is de-energized. Eq. (7) shows the logic model of block TOF.

$$\text{If } (\text{In\_1} = 0 \,\&\, \&\text{ET} \leq \text{PT}), \text{ then Out1} = 0 \tag{7}$$

Restart condition: If In_1 = 1, then ET = 0 and Out1 = 1.

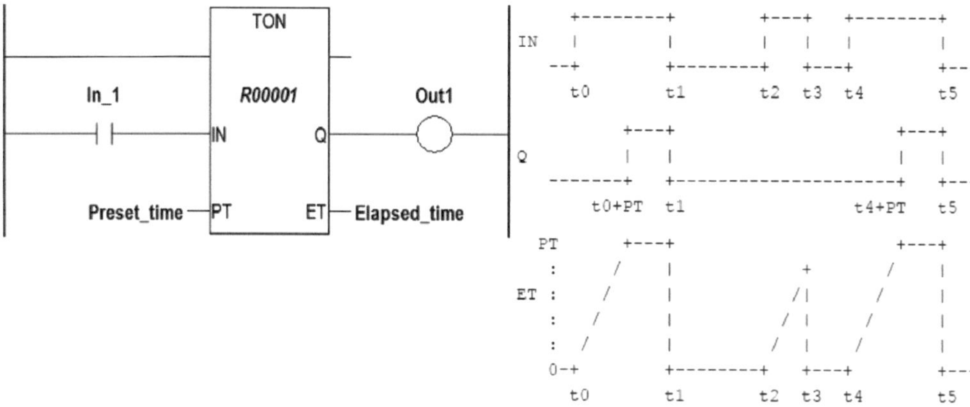

Figure 6. On-delay timer.

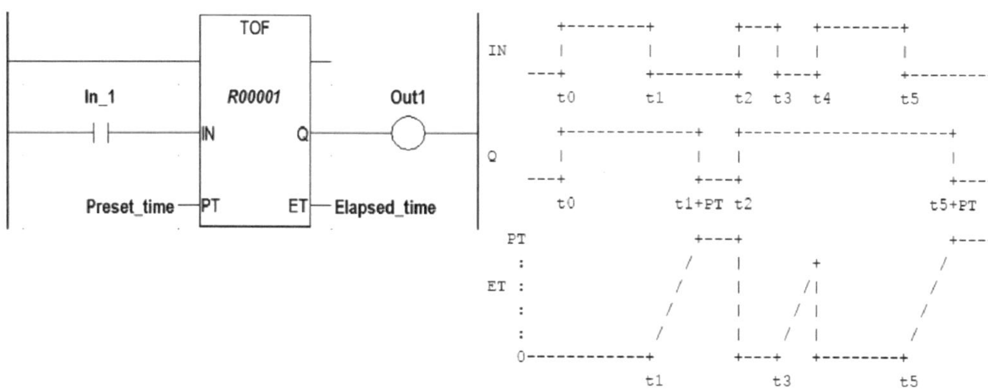

Figure 7. Off-delay timer.

Figure 8 shows two counter function blocks: (1) up-counter and (2) down-counter. In both blocks, the contact In_1 is the pulse to counter, that is and positive transition is detected; if the contact In_2 allows electric energy flow, then Out1 coil is de-energized, and the *Current_value* variable is set to zero in up-counter and to *Preset_value* in down-counter. In up-counter, if *Current_value* is equal or greater than *Preset_value,* then Out1 coil is energized. In down counter, if *Current_value* is equal to zero, then Out1 coil is energized. Eqs. (8) and (9) are logic models of the counters, respectively.

$$\text{if In\_1 }(\uparrow)\text{, then CV} = \text{CV} + 1;\text{ if }(\text{In\_2} = 0\,\&\&\,CV \geq PV)\text{, } then\ Out1 = 1 \tag{8}$$

Restart condition: If In_2 = 1, then CV = 0

$$\text{if In\_1 }(\uparrow)\text{, then CV} = \text{CV} - 1;\text{ if }(\text{In\_2} = 0\,\&\&\,CV \leq 0)\text{, then Out1} = 1 \tag{9}$$

Restart condition: If In_2 = 1, then CV = PV.

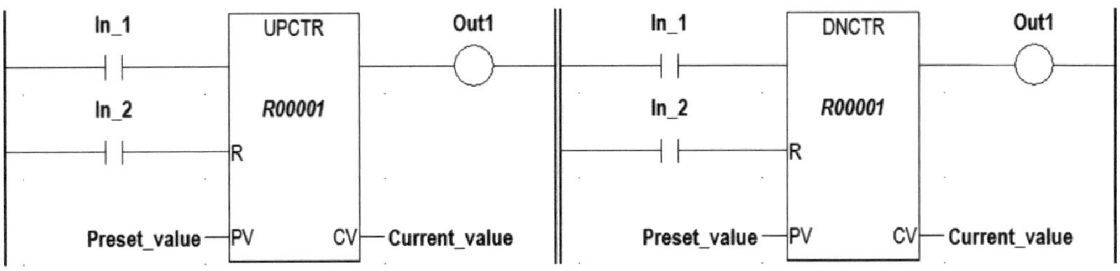

Figure 8. Counters. a) Counter up, b) Counter down.

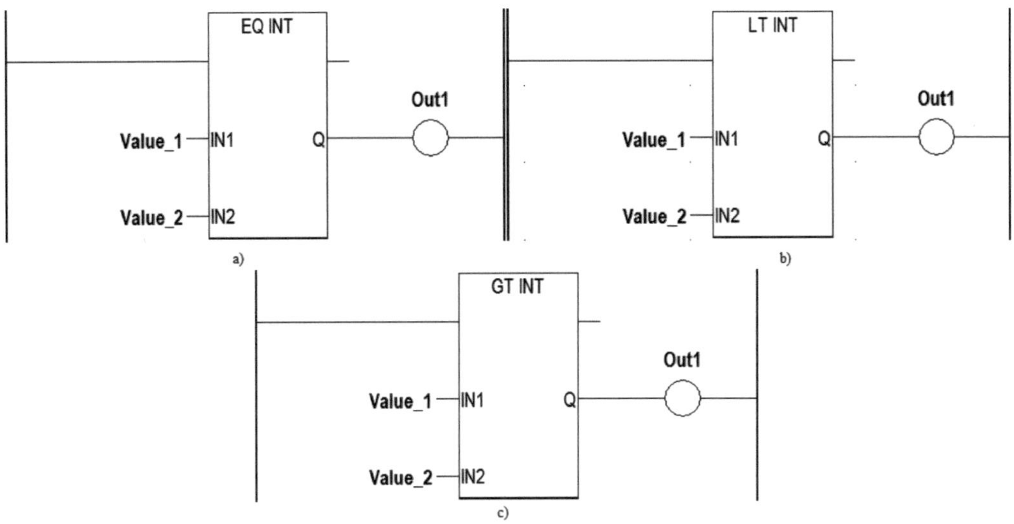

Figure 9. Mathematical comparisons. a) Relation, equal to, b) Relation, lower than, c) Relation, greater than.

Figure 9 shows the standard comparison function blocks: a) equal to, b) lower than and c) greater than. In all the blocks, two analog signals are compared, and depending on result is energized or de-energized Out1 coil. The logic models, respectively, are specified in Eqs. (10–12).

$$\text{If Value\_1} = \text{Value\_2, then Out1} = 1 \tag{10}$$

$$\text{If Value\_1} < \text{Value\_2, then Out1} = 1 \tag{11}$$

$$\text{If Value\_1} > Value\_2,\ then\ Out1 = 1 \tag{12}$$

3. Model of control lines in PN

In this section, the bases of the PN theory are indicated, and the discrete-LDPN network, which is the basis for generating the PN structures of the control lines in LD, is defined.

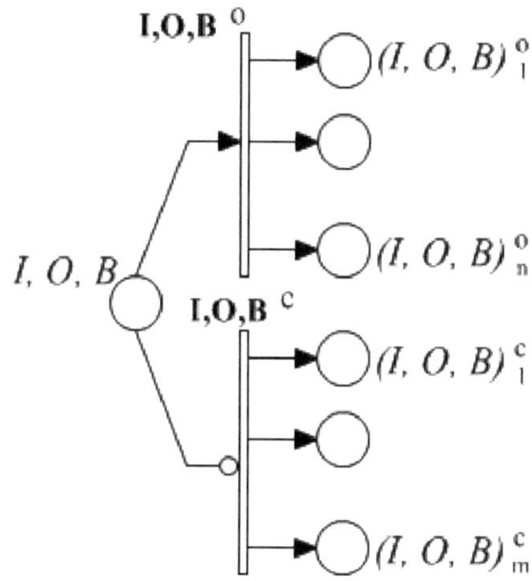

Figure 10. Distribution of discrete signals in PN.

Likewise, the conditions for marking places and triggering transitions are described to model the cyclical evaluation behavior of the control algorithm in PLC.

3.1. Petri nets

PN are a graphic and mathematic tool mean to modeling DES behavior. Graphically, a PN uses circles in order to represent places, rectangles to represent transitions and arcs with arrow or circle to link the inputs and output places with a transition. The relation between places and transition can be represented mathematically by means of an incidence matrix. For a PN with n transitions and m places, its incidence matrix $A = [a_{ij}]$ is an integer number matrix representing the weighting of the input and output arcs; a_{ij}^{+} represents the weighting of output arcs from transitions and a_{ij}^{-} represents input arcs to transitions. Eq. (13) represents how the incidence matrix values are obtained.

$$a_{ij} = a_{ij}^{+} - a_{ij}^{-} \tag{13}$$

To model the dynamic behavior of DES, PN has the state equation, which shows the marking in the net sequentially from initial marking M_{k-1} and when applying a firing vector u_k to the transpose of the respective incidence matrix A^T, respectively. Eq. (14) shows the relationship between them.

$$M_k = M_{k-1} + A^T u_k \tag{14}$$

3.2. LDPN: Discrete event systems

In an LD control algorithm, a discrete signal can have n contacts normally open and m contacts normally closed. The work in [12] shows a representation of discrete signals used in LD to PN, which is the base of conversion of control lines that have both discrete inputs and outputs. On the other hand, evaluation of control algorithm in PLC is cyclical, which generates two important conditions to consider in the PN model; the cyclical evaluation in PN would generate accumulation of marks in the places, and in function of the logic, marking and consuming of theses in places that represent coils in the LD. This last condition is also necessary to restore the information of places in PN that represent physical analog signals or memory registers. **Figure 10** shows the distribution of discrete signals in PN, and Eq. (15) its interpretation. Only one transition can be enabled at a time; if the input place does not have a mark, then the transition $(\mathbf{I}, \mathbf{O}, \mathbf{B})^c$ is enabled for inhibitor arc. Eq. (16) is the generalization of the marking of I, O y B places.

$$I_i = \{I_n^o \cup I_m^c\}; O_o = \{O_n^o \cup O_m^c\}; B_b = \{B_n^o \cup B_m^c\} \tag{15}$$

where the subscripts n and m are not necessarily equal.

$$M(I, O, B) = \begin{bmatrix} 0 \\ 1 \end{bmatrix}, then \left\{ \begin{matrix} M(I, O, B)^o = 0 \; and \; M(I, O, B)^c = 1 \\ M(I, O, B)^o = 1 \; and \; M(I, O, B)^c = 0 \end{matrix} \right\} \tag{16}$$

Considering symbols of [3], for a pre-set and post-set of places, are defined:

$*t = \{p : (p, t) \in F\}$, the set of input places of t.

$*t = \{p : (t, p) \in F\}$, the set of output places of t.

For tokens accumulation problem in input places, the Eqs. (17) and (18) are proposed. Both equations are is in function of the marking of inputs places and of output place. Eq. (17) is for structures with logic AND, and Eq. (18) for logic OR.

$$(O, B)(t*) = \left\{ \prod M(*t) = 1 \,\&\,\& \,(O, B)(t*) = 0 \right\} \tag{17}$$

$$(O, B)(t*) = \left\{ \sum M(*t) = 1 \,\&\,\& \,(O, B)(t*) = 0 \right\} \tag{18}$$

In the same way, to consume token in output place and restoring conditions of PN structures, Eqs. (19) and (20) are defined, which are in function of both marking input places and output places.

$$RC(t*) = \left\{ \prod M(*t) = 0 \,\&\,\& (O, B)(t*) = 1 \right\} \tag{19}$$

$$RC(t*) = \left\{ \sum M(*t) = 0 \,\&\,\& (O, B)(t*) = 1 \right\} \tag{20}$$

From the above, the ladder diagram Petri net: discrete event systems is defined as it is shown in **Table 1**.

Eq. 16 to distribution of signals, Eqs. 17 and 18 to accumulate tokens and Eqs. 19 and 20 to restart conditions should be evaluated after each marking of the net M_{k+1} to update the marking of LDPN and simulate cycled behavior of PLC. Marking of input places I is in function of discrete sensors states.

LDPN considers the following transition rules to dynamic behavior:

- In initial conditions of LDPN, inhibitor arcs enable transitions and put token in its output places O and/or B in PN model with both inputs and outputs discrete. In AI places restart condition of data.
- All output places (O and B) of the PN model are binary, only one can token.
- All transitions enabled should be fired in one some evaluation. To PN model with both inputs and outputs discrete, transition fired T consume unique token $\mathbf{W}(P, T) = 1$ of each input place **P** of **T** and put to unique token $\mathbf{W}(T, P) = 1$ to each output place **T** of **P**. For PN model with some analog input place and output place discrete, the transition **T** should be fired when it satisfies the respective condition (*if - then*) and put to unique token in each output place **T** of **P**.
- To update, marking should be applied Eqs. 16–20.

3.3. Model of control lines both discrete inputs and outputs

Figure 11 shows the PN model of logic AND, if input places I_1^o, O_3^c y B_2^o have a token, then $\mathbf{L_1}$ transition is enabled. The $\mathbf{L_1}$ firing puts a token at place $O_{1.}$ When are updates the marking of input places, the Eq. (17) disables the $\mathbf{L_1}$ transition, avoiding a token more in output place O_1.

A Discrete-LDPN is a 5-tuple (**P, T, W, F, $\mathbf{M_0}$**), where:

$\mathbf{P} = \{I \cup O \cup B \cup AI \cup AR \cup RC\}$ is a finite set of places, where:
$I = \{I_1, I_2, \ldots, I_i\}$ is a finite set of places that represent discrete physical inputs,
$O = \{O_1, O_2, \ldots, O_o\}$ is a finite set of places that represent discrete physical outputs,
$B = \{B_1, B_2, \ldots, B_b\}$ is a finite set of places that represent discrete memory signals,
$AI = \{AI_1, AI_2, \ldots, AI_{ai}\}$ is a finite set of places that represent analog physical inputs,
$AR = \{AR_1, ARI_2, \ldots, AR_{ar}\}$ is a finite set of places that represent analog memory signals,
$RC = \{RC_1, RC_2, \ldots, RC_{rc}\}$ is a finite set of places to restart condition of the nets and its marking it in function of the states of inputs and outputs of control line type.

$\mathbf{T} = \left\{\mathbf{I}^{c|o},, \mathbf{O}^{c|o},, \mathbf{B}^{c|o}, \mathbf{L}, \mathbf{AI}, \mathbf{RC}\right\}$ is a finite set of transitions, where:
$\mathbf{I}^{c|o} = \left\{\mathbf{I}_1^{c|o}, \mathbf{I}_2^{c|o}, \ldots, \mathbf{I}_i^{c|o}\right\}$ is a finite set of transitions that have discrete physical inputs,
$\mathbf{O}^{c|o} = \left\{\mathbf{O}_1^{c|o}, \mathbf{O}_2^{c|o}, \ldots, \mathbf{O}_o^{c|o}\right\}$ is a finite set of transitions that have discrete physical outputs,
$\mathbf{B}^{c|o} = \left\{\mathbf{B}_1^{c|o}, \mathbf{B}_2^{c|o}, \ldots, \mathbf{B}_b^{c|o}\right\}$ is a finite set of transitions that have discrete memory signals,
$\mathbf{L} = \{\mathbf{L}_1, \mathbf{L}_2, \ldots, \mathbf{L}_l\}$ is a finite set of transitions that may have places of discrete signals,
$\mathbf{AI} = \{\mathbf{AI}_1, \mathbf{AI}_2, \ldots, \mathbf{AI}_{ai}\}$ is a finite set of transitions that can have discrete and/or analog signals, its fire condition it in function of mathematics or logics restrictions.
$\mathbf{RC} = \{\mathbf{RC}_1, \mathbf{RC}_2, \ldots, \mathbf{RC}_{rc}\}$ is a finite set of transitions that have input place RC to restart condition of PN structure.

$\mathbf{F} \subseteq (\mathrm{P} \times \mathrm{T}) \cup (\mathrm{T} \times \mathrm{P})$ is a set of arcs.

$\mathbf{W} = \mathbf{F} \rightarrow \{1\}$, all weights of the arcs are equal to 1.

$$\mathbf{M}_0 = \begin{cases} \mathrm{P} \rightarrow \{0, 1\}, & \text{discrete signal.} \\ \mathrm{P} \rightarrow \{\mathrm{Z}\ (16\ \text{bit integer})\}, & \text{analog signal.} \end{cases}$$

Table 1. Definition of LDPN: Discrete event systems.

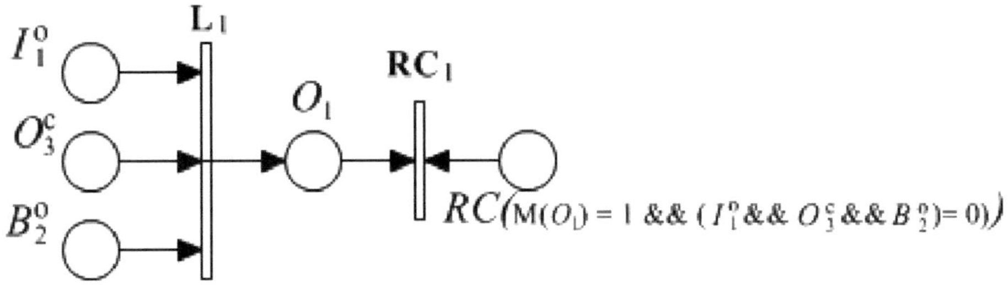

Figure 11. PN model of logic AND.

By Eq. (19), the marking of place *RC* is in function of both marking input places and output place.

Figure 12 shows the PN model of logic OR, if any input places I_1^c, O_5^o *y* B_2^o have a token, then $\mathbf{L_1}$, $\mathbf{L_2}$ or $\mathbf{L_3}$ transition is enabled, respectively. If the transition enabled is fired, then a token is put at place O_1. When are updates the marking of input places, the Eq. (18) disables the $\mathbf{L_1}$, $\mathbf{L_2}$ and $\mathbf{L_3}$ transitions, avoiding a toke more in output place O_1. By Eq. (20), the marking of place *RC* is in function of both marking input places and output place.

Figure 13 shows the PN model of logic AND-OR, output place can get token from L_1 or L_2 transitions, in function of marking of input places I_1^o, O_3^c *y* B_2^o or I_1^o, O_3^c *y* O_7^c have a token, respectively. The $\mathbf{L_1}$ or $\mathbf{L_2}$ firing puts a token at place O_1. When are updates the marking of input places, the Eqs. (17) and (18) disables the $\mathbf{L_1}$ and $\mathbf{L_2}$ transitions, avoiding a token more in output place O_1. The marking of place *RC* is in function of both marking input places of $\mathbf{L_1}$ and $\mathbf{L_2}$ transitions and output place O_1 based on Eqs. (19) and (20) to restart condition.

Figure 14 shows the PN model of logic auto-loop. In this model, it is necessary that $\mathbf{L_1}$ transition to be enabled and fired set a token in the output place O_1, enabling the $\mathbf{O_1^o}$ transition, which consumes the token of O_1 and sets a token in the place O_1^o, enabling the $\mathbf{L_2}$ and holding a token in O_1. The restart condition of the model auto-loop is in function of the Eqs. (19) and (20).

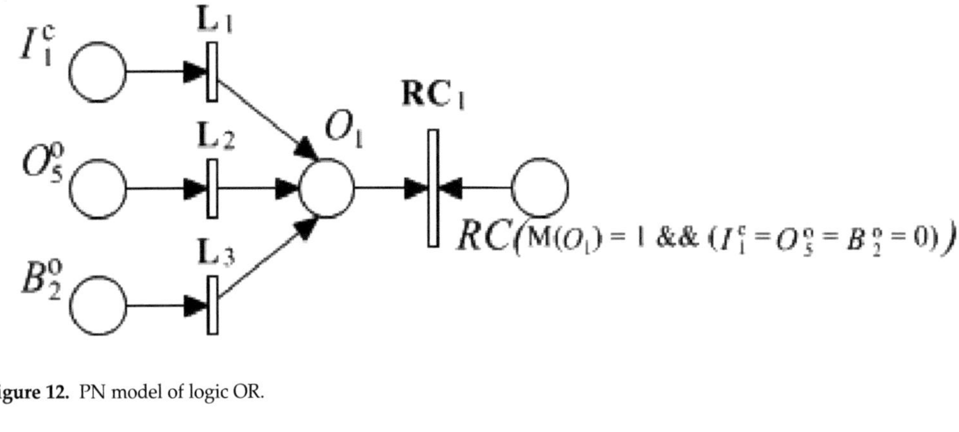

Figure 12. PN model of logic OR.

http://dx.doi.org/10.5772/intechopen.75753

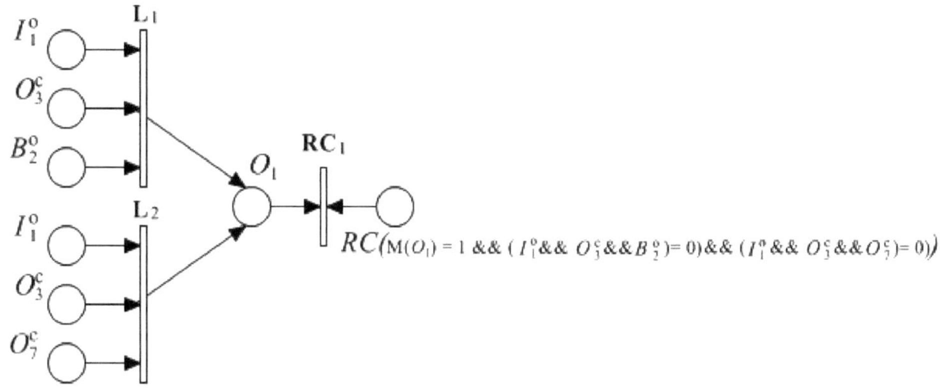

Figure 13. PN model of logic AND–OR.

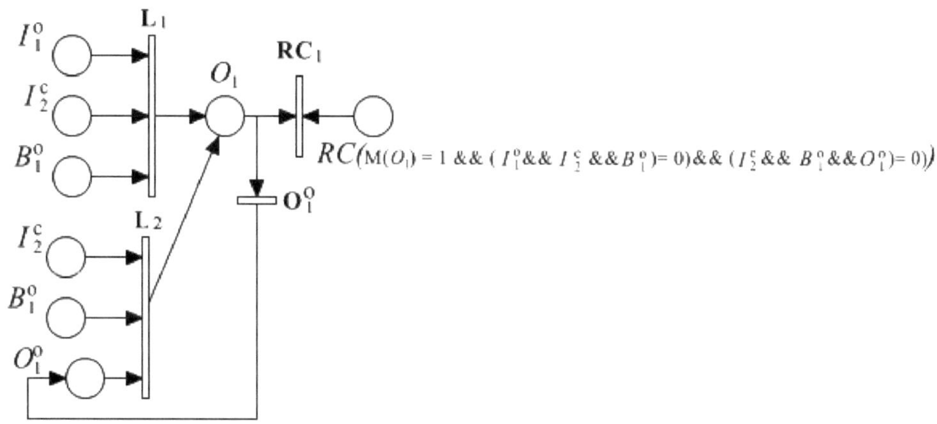

Figure 14. PN model of logic auto-loop.

Figure 15 shows the PN model of logic interlocking. Both places O1 and O2 enable the $\mathbf{O_1^c}$ and $\mathbf{O_2^c}$ transitions by the inhibitor arcs, placing a token in input places O_1^c and O_2^c of $\mathbf{L_1}$ and $\mathbf{L_2}$ transitions, respectively. If $\mathbf{L_1}$ or $\mathbf{L_2}$ transition is firing first disables the other transition by the inhibitor arc. The restart condition places RC_1 and RC_2 are in function of Eq. (19).

3.4. Model of control lines with analog inputs and output discrete

Figure 16 shows the PN model of on-delay timer. The BT and PT are variables to determine base time and preset time, respectively. The marking of the place AI_2 is a data analog to store the sum ET = ET + BT. The marking of the place O_1 is in function of firing of the AI_2 transition, which depends on the condition $if(\, I_1 = 1 \,\&\,\& \, ET \geq PT)$. To restart condition of the places O_1 and AI_2 are in function of I_1^c.

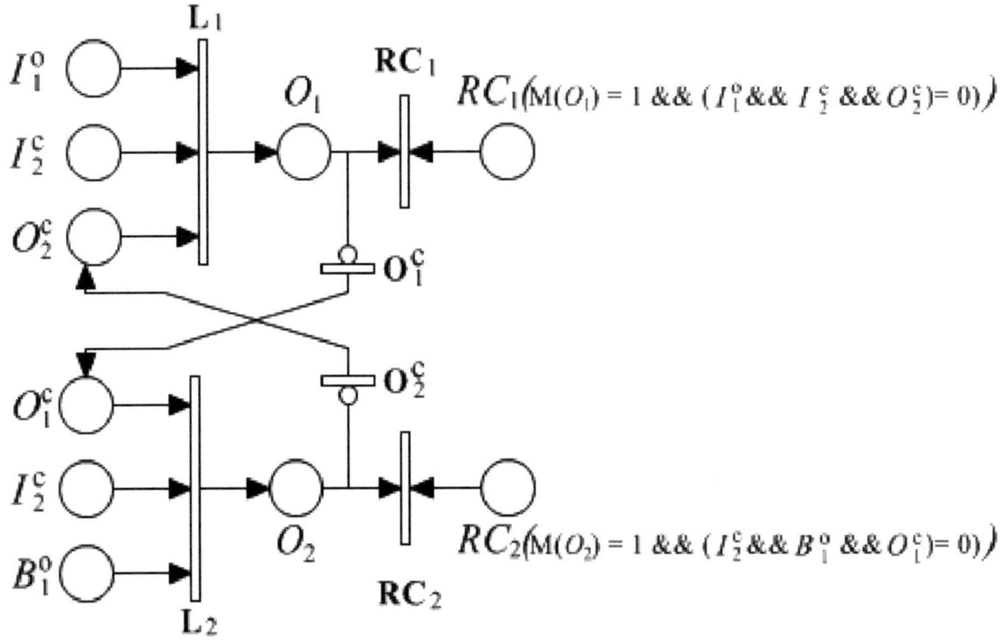

Figure 15. PN model of logic interlocking.

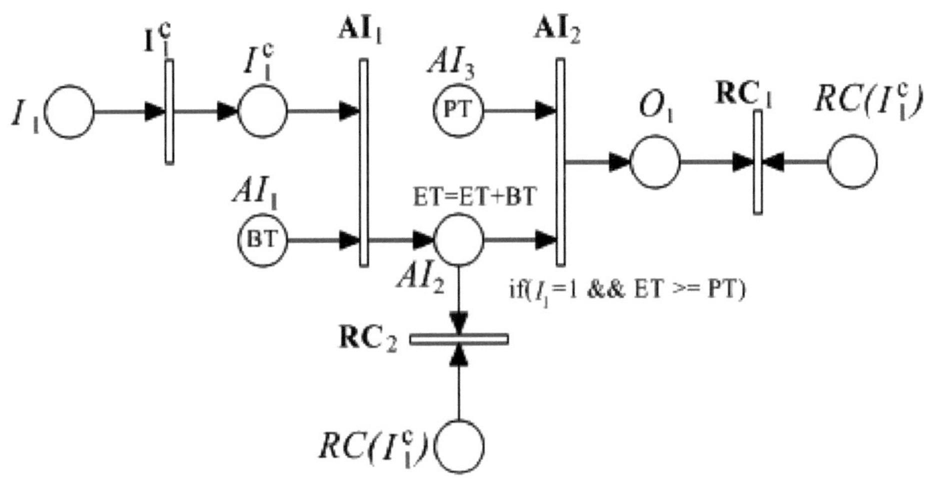

Figure 16. PN model of on-delay timer.

Figure 17 shows the PN model of logic off-delay timer. The place to restart condition RC and fire of $\mathbf{RC_1}$ is putting a token in output place O_1 that is the initial condition of the structure PN. When the $\mathbf{I_1^c}$ transition is fired put a token in place I_1^c, which enables $\mathbf{AI_1}$ transition to allow the sum of time $ET = ET + BT$. The fire of $\mathbf{AI_2}$ transition is in function of $if(I_1 = 0\,\&\&\,ET \leq PT)$, if

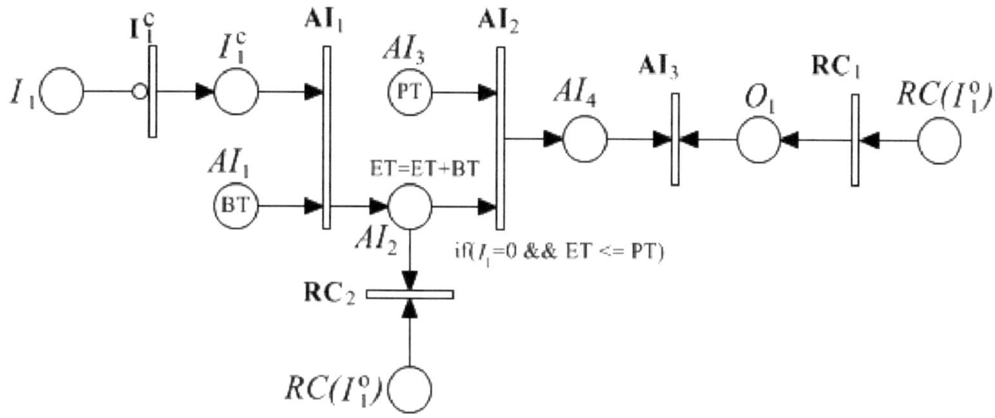

Figure 17. PN model of logic off-delay timer.

it is fired, then put a token in place AI_4, which enables $\mathbf{AI_3}$ transition and consumes the token of place O_1.

Figure 18 shows the PN model of logic up-counter and **Figure 19** to PN model of logic down-counter. In both models, the marking of the place I_1^o is in function of $M(I_1^o) = \{M(I_1(*t)) = 0 \,\&\&\, M(I_1(t*)) = 0\}$, which is to detect a positive transition in the marking, respectively. In place AI_1 are added the tokens (positive transition), *if* $(CV \geq PV)$ then $\mathbf{AI_1}$ transition is enabled, and its fire put a token in place O_1. If the place $RC(I_2^o)$ has a token, then it is consumed the token of the place O1 and $CV = 0$.

Figure 19 shows the PN model of logic down-counter, which has similar behavior to up-counter, just that if one token in place I_1^o consume one token of place AI_1, *if* $(CV \leq 0)$, then the fire of $\mathbf{AI_1}$ transition puts a token in place O_1.

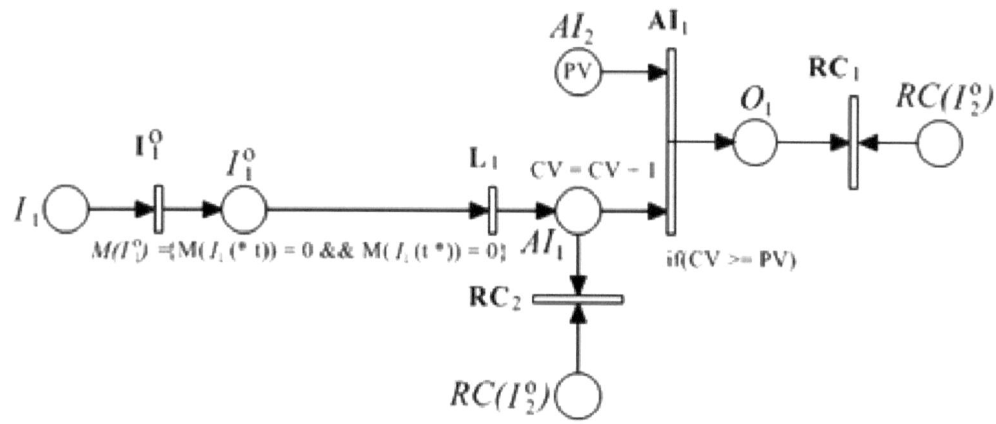

Figure 18. PN model of logic up-counter.

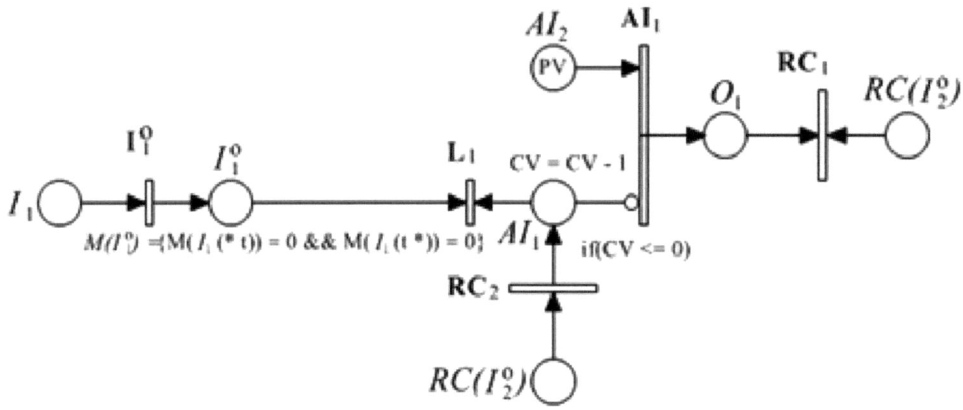

Figure 19. PN model of logic down-counter.

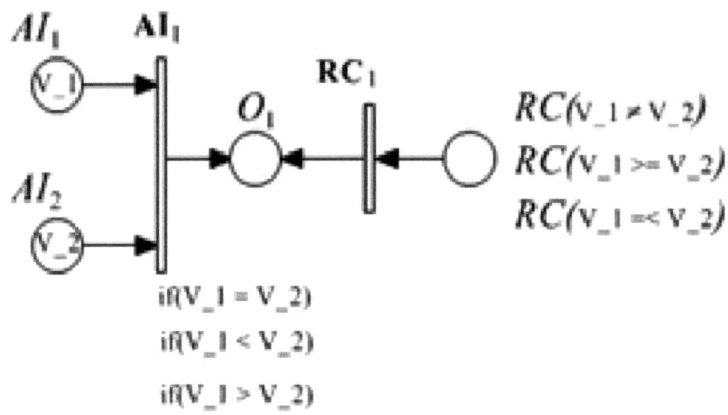

Figure 20. PN model of logic of comparisons.

Figure 20 shows the PN model of logic of comparisons of two analog places. Enabling and firing of $\mathbf{AI_1}$ is in function of $if(V_1 = V_2); if(V_1 < V_2); if(V_1 > V_2);$ according to the comparison. Similarly, the marking of place RC is in function of $RC(V_1 \neq V_2); RC(V_1 \geq V_2);$ $RC(V_1 \leq V_2)$, respectively.

4. Example

Figure 21 shows the control algorithm in LD of run of three motors sequentially [14]. The Start and Stop signals are physical inputs of type pushbutton. The Motor_1, Motor_2 and Motor_3 coils are physical outputs. The IR1, IR2 and IR3 variables are bits of memory. The first control line is logic of auto-loop, if Start variable is equal to one, then, the IR1 coil is energized and so it is hold by the contact IR1. It is also energized the Motor_1 coil, and the timer T1 and T2 begin counting time. In T1, if $ET \geq PT$, then, the IR2 and Motor_2 coils are energized. In T2, if $ET \geq PT$,

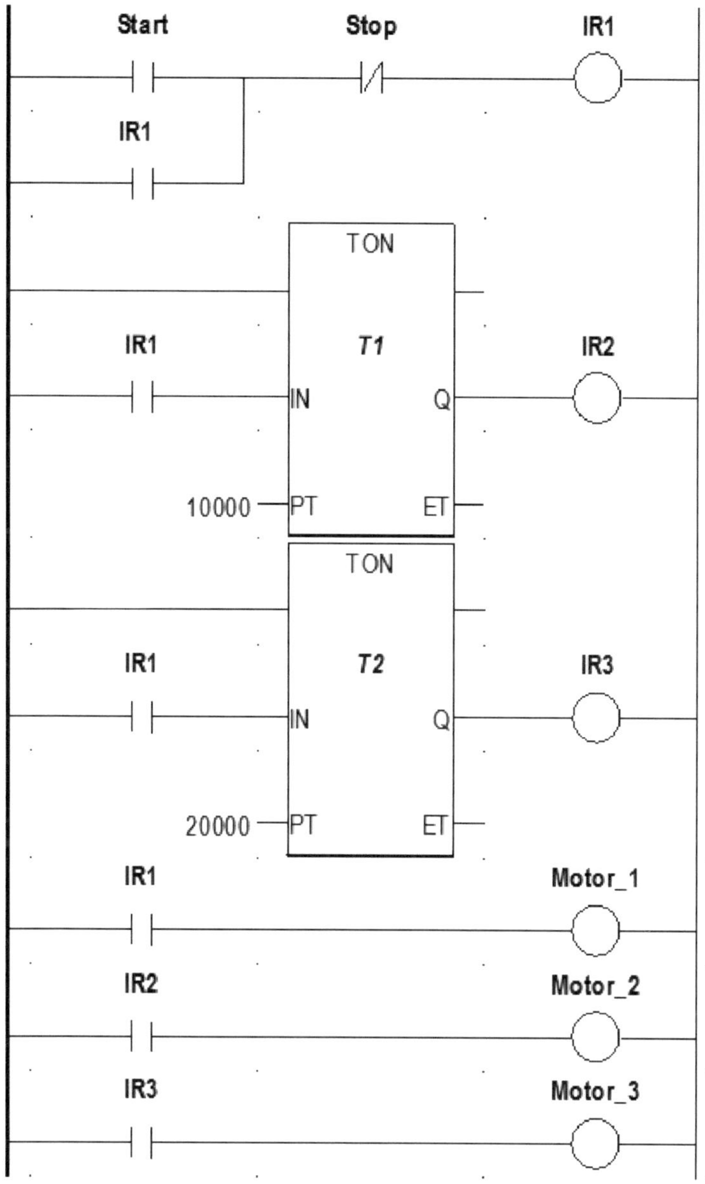

Figure 21. Run of three motors sequentially.

then, the IR3 and Motor_3 coils are energized. If Stop = 1, then, the IR1 coil is de-energized, and are restart conditions of the control algorithm. **Table 2** shows the equivalence of signals in function of the definition LDPN.

Figure 22 shows the LDPN to control algorithm of run of three motors sequentially. Restart conditions of the output places are in function of B_1^c by Eq. 19. The restarting condition of place B_1 is in function of input places I_1 and I_2 by Eqs. 19 and 20. Restart condition places RC_2 to RC_6

LD	LDPN
Start	I_1
Stop	I_2
Motor_1	O_1
Motor_2	O_2
Motor_3	O_3
IR1	B_1
IR2	B_2
IR3	B_3

Table 2. Equivalence of signals of control algorithm in LDPN.

are in function of the marking B_1^o, which are connected from $\mathbf{B_1^c}$. For complex control algorithms implies a larger graphic LDPN, it is advisable to indicate the marking function of the places *RC*. Eq. 21 shows the incidence matrix of the PN model respectively, where the conditioning *if-then* of transitions for reasons of space, which are indicated on corresponding figures, is omitted.

$$
\mathrm{A_{ij}} =
\left[\begin{array}{cccccccccccccccccccccccccc}
 & I_1 & I_1^o & I_2 & I_2^c & I_2^c & B_1 & B_1^o & B_1^o & B_1^o & B_1^o & AI_1 & AI_2 & AI_3 & AI_4 & AI_5 & AI_6 & O_1 & O_2 & O_3 & RC_1 & RC_2 & RC_3 & RC_4 & RC_5 & RC_6 \\
\mathbf{I_1^o} & -1 & 1 & & & & & & & & & & & & & & & & & & & & & & & \\
\mathbf{I_2^c} & & & -1 & 1 & 1 & & & & & & & & & & & & & & & & & & & & \\
\mathbf{B_1^o} & & & & & & -1 & 1 & 1 & 1 & 1 & & & & & & & & & & & & & & & \\
\mathbf{B_1^c} & & & & & & -1 & & & & & & & & & & & & & & & 1 & 1 & 1 & 1 & 1 \\
\mathbf{L_1} & & -1 & & -1 & & 1 & & & & & & & & & & & & & & & & & & & \\
\mathbf{L_2} & & & & & -1 & 1 & -1 & & & & & & & & & & & & & & & & & & \\
\mathbf{L_3} & & & & & & & & -1 & & & & & & & & & 1 & & & & & & & & \\
\mathbf{AI_1} & & & & & & & & & -1 & & \mathrm{BT=1ms} & \mathrm{ET+BT} & & & & & & & & & & & & & \\
\mathbf{AI_2} & & & & & & & & & & & & \mathrm{ET+BT} & \mathrm{PT}=10^4 & & & & & 1 & & & & & & & \\
\mathbf{AI_3} & & & & & & & & & -1 & & & & & \mathrm{BT=1ms} & \mathrm{ET+BT} & & & & & & & & & & \\
\mathbf{AI_4} & & & & & & & & & & & & & & & \mathrm{ET+BT} & \mathrm{PT}=2*10^4 & & & 1 & & & & & & \\
\mathbf{RC_1} & & & & & & -1 & & & & & & & & & & & & & & -1 & & & & & \\
\mathbf{RC_2} & & & & & & & & & & & & & & & & & -1 & & & & -1 & & & & \\
\mathbf{RC_3} & & & & & & & & & & & & & & & & & & -1 & & & & -1 & & & \\
\mathbf{RC_4} & & & & & & & & & & & & \mathrm{ET} & & & & & & & & & & & -1 & & \\
\mathbf{RC_5} & & & & & & & & & & & & & & & & & & & -1 & & & & & -1 & \\
\mathbf{RC_6} & & & & & & & & & & & & & & & \mathrm{ET} & & & & & & & & & & -1
\end{array}\right]
$$

The dynamic behavior of the PN model by run of three motors sequentially is described by the following marking. Fired the transitions with inhibitor arcs, initial marking M_0 is:

$$
M_0 = \begin{bmatrix} I_1 & I_1^o & I_2 & I_2^c & I_2^c & B_1 & B_1^o & B_1^o & B_1^o & B_1^o & AI_1 & AI_2 & AI_3 & AI_4 & AI_5 & AI_6 & O_1 & O_2 & O_3 & RC_1 & RC_2 & RC_3 & RC_4 & RC_5 & RC_6 \\ 0 & 0 & 0 & 1 & 1 & 0 & 0 & 0 & 0 & 0 & 1ms & 0 & 10^4 & 1ms & 0 & 2*10^4 & 0 & 0 & 0 & 0 & 1 & 1 & 1 & 1 & 1 \end{bmatrix} \tag{21}
$$

If place I_1 has token, which enable the $\mathbf{I_1^o}$ transition, its fire puts a token in the place I_1^o, which enabled the $\mathbf{L_1}$ transition, its fire puts a token in the place B_1. In these conditions, by Eq. (16), the tokens in places of restarting conditions are consumed; the marking corresponding of LDPN is shown in Eq. (23).

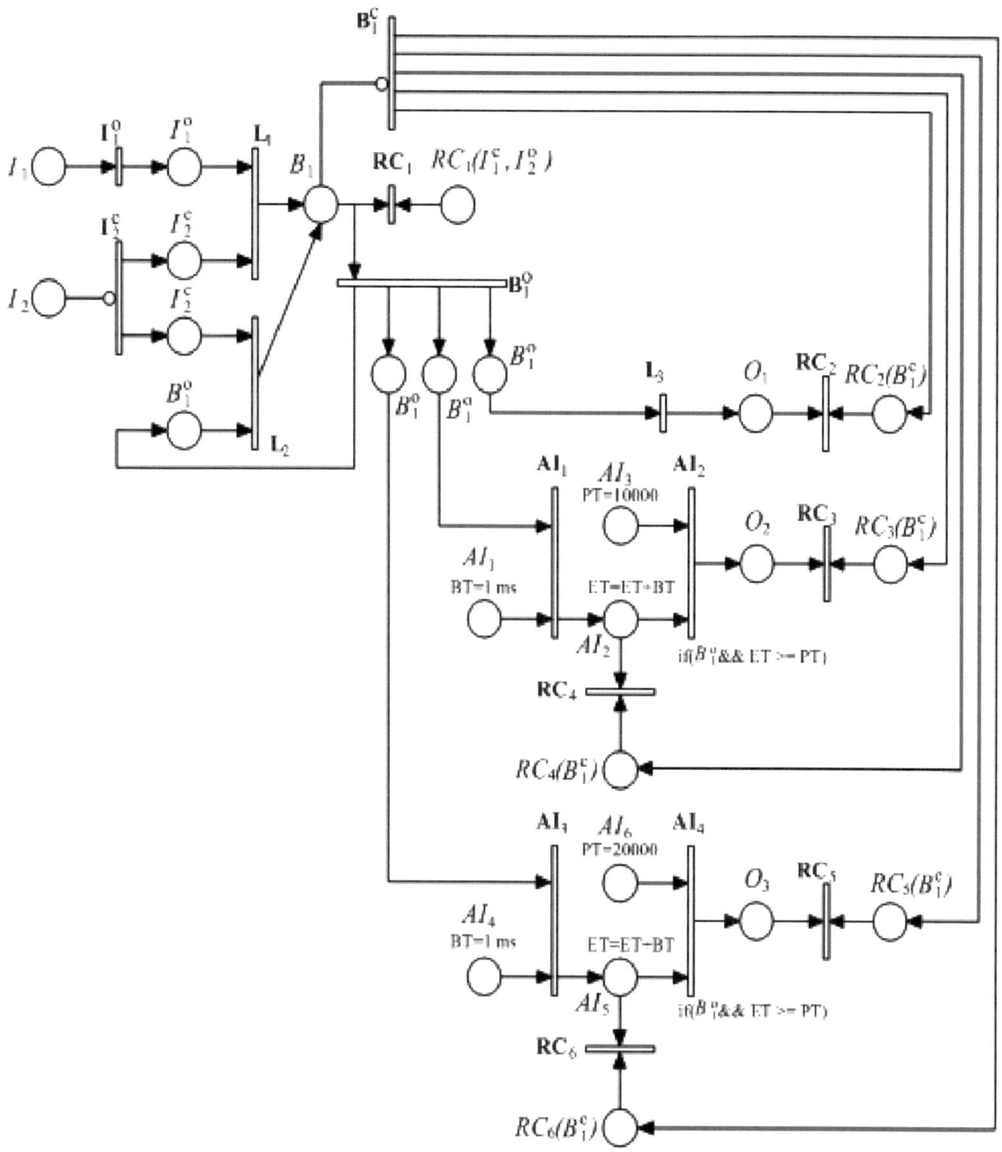

Figure 22. PN model of run of three motors sequentially.

$$M_1 = \begin{bmatrix} I_1 & I_1^o & I_2 & I_2^c & I_2^c & B_1 & B_1^o & B_1^o & B_1^o & B_1^o & AI_1 & AI_2 & AI_3 & AI_4 & AI_5 & AI_6 & O_1 & O_2 & O_3 & RC_1 & RC_2 & RC_3 & RC_4 & RC_5 & RC_6 \\ 0 & 0 & 0 & 0 & 1 & 1 & 0 & 0 & 0 & 0 & 1ms & 0 & 10^4 & 1ms & 0 & 2*10^4 & 0 & 0 & 0 & 0 & 0 & 0 & 0 & 0 & 0 \end{bmatrix} \tag{22}$$

In these conditions, the $\mathbf{B_1^o}$ transition is enabled, its fire puts a token in four places B_1^o, this enables $\mathbf{L_2}$ transition, its fire puts a new token in the place B_1, it disables the fire of $\mathbf{L_1}$ and $\mathbf{L_2}$ transitions by Eqs. (17) and (18). Another place B_1^o enables $\mathbf{L_3}$ transition; its fire puts a token in the place O_1. The others two places B_1^o enable $\mathbf{AI_1}$ and $\mathbf{AI_3}$ transition to add the base time, respectively. Eq. (24) shows these conditions of LDPN, besides the update marking.

$$M_1 = \begin{bmatrix} I_1 & I_1^o & I_2 & I_2^c & I_2^c & B_1 & B_1^o & B_1^o & B_1^o & B_1^o & AI_1 & AI_2 & AI_3 & AI_4 & AI_5 & AI_6 & O_1 & O_2 & O_3 & RC_1 & RC_2 & RC_3 & RC_4 & RC_5 & RC_6 \\ 0 & 0 & 0 & 1 & 1 & 1 & 1 & 1 & 1 & 1 & 1ms & Add & 10^4 & 1ms & Add & 2*10^4 & 1 & 0 & 0 & 0 & 0 & 0 & 0 & 0 & 0 \end{bmatrix} \tag{23}$$

In these conditions, if $ET \geq PT$, in both $\mathbf{AI_2}$ and $\mathbf{AI_4}$ transitions put a token in places O_2 and O_3, respectively. When place I_2^o has a token enabling the $\mathbf{RC_1}$ transition, its fire consumes a token in the place B_1, restarting condition in LDPN.

5. Conclusions

There are two types of control lines for discrete event systems: those with discrete inputs and outputs, and those with analog inputs and discrete output. Twelve logics that were analyzed and converted into Petri network models.

For dynamic behavior of the PN model proposed, constraints and equations for marking places and firing transitions are indicated to consider the problems of mark accumulation and the restarting condition of the structure PN.

LDPN to discrete event systems allow to model control lines used in LD language, and consequently, control algorithms development in LD, supporting that these are safe and reliable.

Each PN model is independent and can be interconnected in function of the control logic, as well as, the number of PN model that is needed can be integrated.

Author details

José Carlos Quezada Quezada[1]*, Ernesto Flores García[1], Joselito Medina Marín[2], Jorge Bautista López[3] and Víctor Quezada Aguilar[1]

*Address all correspondence to: jcarlos@uaeh.edu.mx

1 High Education School Tizayuca, Autonomous University of Hidalgo State, Mexico

2 Advanced Research Center in Industrial Engineering, Autonomous University of Hidalgo State, Mexico

3 Campus Zusmpango, Autonomous University of Mexico State, Mexico

References

[1] International Electrotechnical Commission, IEC 61131-3: Programmable Controllers: Programming Languages, International standard, 2nd ed, 2003

[2] Korotkin S, Zaidner G, Cohen B, Ellenbogen A, Arad M, Cohen Y. A petri net formal design methodology for discrete-event control of industrial automated systems, IEEE 26-th convention of electrical and electronics engineers in Israel; 2010. pp. 431-435. DOI: 10.1109/EEEI.2010.5662187

[3] John K-H, Tiegelkamp M. IEC 61131–3: Programming Industrial Automation Systems. 2nd ed. Springer; 2010

[4] Murata. Petri Nets: Properties, analysis and applications. Proceedings of the IEEE. 1989. pp. 541-580. DOI: 10.1109/5.24143

[5] Luo J, Zhang Q, Chen X, Zhou MC. Modeling and Race Detection of Ladder Diagrams via Ordinary Petri Nets. IEEE Transactions on Systems, Man and Cybernetics. DOI: 10.1109/TSMC.2016.2647219

[6] Năvrăpescu V, Deaconu I-D, Chirilă A-I, Deaconu A-S. Petri Net versus Ladder Diagram for Controlling a Process Automation. The 8th International symposium on advanced topics in electrical engineering. May 23–25, Bucharest, Romania, 2013. DOI: 10.1109/ATEE.2013.6563402

[7] Xuekum C, Lilian L, Pengfei Q. Method For Translating Ladder Diagram To Ordinary Petri Nets. 51st IEEE conference on decision and control; 2012. pp. 6716-6721. DOI: 10.1109/CDC.2012.6426901

[8] Zhang H, Jiang Y, Hung WN, Yang G, Gu M, Sun J. New strategies for reliability analysis of programmable logic controllers. Mathematical and Computer Modeling. 2012;**55**(7/8): 1916-1931. DOI: 10.1016/j.mcm.2011.11.050

[9] da Silva Oliveira EA, da Silva LD, Gorgonio K, Perkusich A, Martins AF. 9th IEEE international conference onObtaining formal models from ladder diagrams, industrial informatics (INDIN), 26-29 July, 2011; p. 796-801. DOI: 10.1109/INDIN.2011.6034994

[10] Lee J, Lee JS. Conversion of ladder diagram to petri net using module synthesis technique. International Journal of Modeling and Simulation. 2009;**29**(1):79-88. DOI: 10.1080/02286203.2009.11442513

[11] Grobelna I, Grobelny M, Adamski M. Petri Nets and Activity Diagrams in Logic Controller Specification—Transformation and Verification. 17th International Conference Mixed Design of Integrated Circuits and Systems, Wroclaw, Poland; 2010. pp. 607-612

[12] Quezada JC, Medina J, Flores E, Seck Tuoh JC, Solís AE. Simulation and validation of diagram ladder – Petri net. International Journal Advance Manufacturing Technology. 2017;**88**:1393-1405. DOI: 10.1007/s00170-016-8638-9

[13] Quezada JC, Medina J, Flores E, Seck Tuoh JC, Hernández N. Formal design methodology for transforming ladder diagram to Petri nets. International Journal Advance Manufacturing Technology. 2014;**73**:821-836. DOI: 10.1007/s00170-014-5715-9

[14] Bolton W. Programmable Logic Controllers. 5th ed. Elsevier Ltd.; 2009. pp. 222-223, ISBN: 978-1-85617-751-1

Chapter 3

Petri Networks in the Planning of Discrete Manufacturing Processes

Roman Stryczek

Additional information is available at the end of the chapter

http://dx.doi.org/10.5772/intechopen.75135

Abstract

This chapter puts forward characteristics of selected issues of manufacturing processes planning using the Petri networks technique. It includes references to the extensive literature concerning the use of Petri networks in computer aided planning of discrete production processes. Diversity of these problems is high as it refers both to the methods of modeling and simulation of the course of manufacturing processes, the issue of optimizing these processes and production systems, representation of knowledge on production parts of equipment and integration of planning and production activities in general. The work puts forward example use of a temporary, priority Petri network for modeling and optimizing production systems and manufacturing operations as well as an example of fuzzy interference using the Petri network mechanism.

Keywords: CAPP, modeling production systems, representation of knowledge, CIM, time-priority Petri nets

1. Introduction

The objective of modeling a manufacturing process is mostly utilitarian with aspects such as support of its planning, ensuring optimizing or providing environment for automation of its planning. Its cognitive aspect is also of importance as building a model forces a planner to track the entire issue of generating a process plan. The main components of a production process description include: a description of the stereo-metric structure (stereo-structure), a description of the time structure (chrono-structure), a specification of processing conditions and a description of the factors of ensuring reliability of processing. Stereo-structure involves characteristics related to dimensional production chains, spatial arrangement and connection with movement of working units of the machine tool. Chrono-structure involves characteristics related to structural components of operation time and the sequence of these components. Chrono-structure of a complex, multi-tool machining processes involves the following components: specification of simple operations completed on individual features; assigning simple operations to complex operations; distribution of operations into complex operations and defining a reasonable sequence of complex operations.

A simple operation understood as processing one feature with one tool is a basic component of the operations chrono-structure. A complex operation means a group of simple operations completed using one tool without replacing it, changing the position of the processed part and without re-mounting, even a partial one. Complex operations are combined in sequences on rising hierarchy levels. Borders of these sequences are set by: change of a tool, change of the parts processing position within the reference frame of the machine tool, re-mounting, change of reference frame and, on a higher level of production process, division into roughing and precise processing.

A manufacturing operation might be defined as a cause-and-effect process including three basic subgroups:

1. Passive (static) including current statuses of features, tools, status of the machine tool's auxiliary assemblies such as tailstock sleeve, steady rest, turntable, pallet changer, position of tool head as well as selected manufacturing datum and clamping methods;
2. Active (dynamic) including operations, changes of tools, movement of auxiliary units, re-clamping, and so on;
3. Decision-making, including events, the nature of which is not temporary, but informative such as releasing the opportunity of changing a tool, forcing a change in the table's position, and so on.

A description of the chrono-structure might be made by defining organization and concurrency relations on sets of components of the aforementioned subgroups. If the first of the subgroups is treated as a set of conditions and the other two as a set of events, and flow relation is described on these two sets, the result is a directed bipartite graph structure—a base of the Petri network.

Topological characteristics of a bipartite directed graph that the Petri network is allowed for modeling various logical, cause and effect, time, attribute, linguistic, semantic, geometrical and other relations. Such relations are considered both at the stage of planning a production process and at the stage of implementing it. Therefore, an interest in the methodology of Petri networks for the purpose of planning manufacturing processes grew as early as in the second half of the 1980s [1]. The first attempts to use the Petri network in planning manufacturing processes were related to connecting the production process plan to conditions resulting from the production department's potential. Therefore, their basic use in construction of machines consisted of modeling of production systems, mostly the flexible manufacturing system (FMS). A comprehensive review of use of Petri networks in planning manufacturing processes by 1992 is put forward in this work [2].

This period was followed by a range of significant applications of Petri networks in modeling discrete manufacturing processes. The team of Kiritsis has showed particular interest in this issue [3–10]. The published works analyzed the opportunities of classifying operations in the process of production, representation of alternative courses of manufacturing process and dynamic planning of processes. The complete approach in the 1990s has also been put forward by the duo of Horvát and Rudas [11–14]. It involved both acquisition of knowledge and modeling the structure of the manufacturing process as well as evaluation of the generated Petri network. The authors aimed at developing a knowledge-based manufacturing process modeling methodology. In the same period, the use of Petri networks was forecast for two purposes: modeling knowledge related to selection and classifying operations and flexible representation of the sequence of operations [15, 16].

Today, it might be stated that this modeling technique has become the most popular one in FMS modeling, scheduling production [17], controlling and management of manufacturing processes in FMS. Stochastic Petri networks are used for considering the random nature of some events [18]. Preventive detection of locks is one of the FMS model's main functions [19].

The issue of flow of objects through the manufacturing process remains in between planning of the manufacturing process and the measures necessary to implement it. The Petri networks methodology is also used here. The work [10] considering the issue of estimating the upper and lower limit of time and cost of completing a production series of a given part in certain workshop conditions is essential here. An opportunity of variant course of individual processes for individual parts of a production batch is assumed. A more efficient model was obtained—compared to the traditional approach based on determining critical path—with regard to the availability of production tools and machinery.

The most expanded Petri network classes in manufacturing are used in the work [20] for the purpose of loading machinery in FMS. Based on of hybrid Petri networks [21], which are fuzzy neural Petri networks, their language was expanded by adding color attributes, inhibitory arcs and time function. According to the authors, the ENhanced Fuzzy Neural Petri Net (ENFNPN) gives extraordinary opportunities of flexible modeling perfectly corresponding to FMS, allowing for modeling and processing knowledge at the expert system level.

Multi-axis machine tools with a larger number of machine units currently play an increasingly significant role in machine processing. The effective use of such machine tools is strictly determined with maximizing concurrency, which is reflected by an optimum chrono-structure of operations. In order to optimize the chrono-structure of a concurrent operation, it is necessary to define subsets of operations that might be conducted simultaneously and to define the optimum relation of preferences, taking into account the division along with a range of various conditions and limitations. It is not a trivial task; hence, it is preferable to use a correctly adapted computer aided process planning (CAPP) class system and planning methodology.

Process specification language (PSL) is the most complex project in modeling and analysis of manufacturing processes using the Petri network technique. The works on this project, which were coordinated by the National Institute of Standards and Technology (NIST) aimed at developing an international standard of a description language for all aspects of manufacturing process completion [6]. Petri network class called the Compact Process Planning net (CPP-net) is the technical representation of PSL concept. Formally, CPP-net is an organized set of four components (P, T, E, M0); it is therefore a base Petri network. A set of P locations in CPP-net includes three subsets: control locations, input locations and locations representing limits. Components of T set represent tasks to be performed, conditioned by the E flow relation. A feasibility graph is developed based on the Petri network. It presents all the possible transition sequences in CPP-net corresponding to possible courses of the process. A large proportion of this course is irrational. Therefore, the subsequent stage includes a heuristic method to eliminate the courses that are not compliant with certain assumptions resulting from correctness of the production process [5]. It allows to significantly reduce the number of possible variants of the course of the process.

2. Generative process planning

Basically, there are two various approaches to plan a process in CAPP systems: generative (GCAPP) and variant (VCAPP). Differences are significant and are also reflected in models of manufacturing processes. An intermediate approach, also called a hybrid one, is also customary, but its description cannot be easily formalized. Generative planning consists of drawing up the process using individual features, to synthesis of tasks on an increasingly higher level: a complex operation, mounting, operation, processing stage, manufacturing process. The generative approach is used if manufactured machine parts significantly differ from each other and groups of parts with manufacturing similarities cannot be identified. The process model in GCAPP is generated gradually as progress is made in planning it. A generated model has a dispersed nature; however, it includes modules with a previously defined structure. It is then necessary to determine the number of modules of a certain type and classify them. **Figure 1** presents the general structure of such models.

The Petri network language class used for building the model has the following form:

$$PN_1 = (P, T, E, I, S, R, M_0), \tag{1}$$

where P: a nonempty, finite set of places, T: a nonempty, finite, disjointed of P set of transitions, $E \subset (P \times T) \cup (T \times P)$: a flow relation, I: a set of inhibitor arcs, S: $T \rightarrow N_0$, a function of time, R: $T \rightarrow [0, 1]$, a priority function, and M_0: $P \rightarrow (0, 1)$, an initial marking.

The conditions of preparing transition t for firing have the following form:

$$M(p) = \begin{cases} 1, \forall p \in {}^{\cdot}t \\ 0, \forall i(p, t) \in \mathrm{I} \end{cases} \tag{2}$$

where $M\ (p)$—a current marking for p place, ${}^{\cdot}t$–t transition input places set, $i(p, t)$—inhibitor arc from p place to t transition.

Eq. (2) applies in cases where only one transition meets the feasibility conditions. If two or more transitions meet the feasibility conditions, only one of them—the highest priority one—is completed. Other transitions are ranked in next iteration of the simulations process.

http://dx.doi.org/10.5772/intechopen.75135

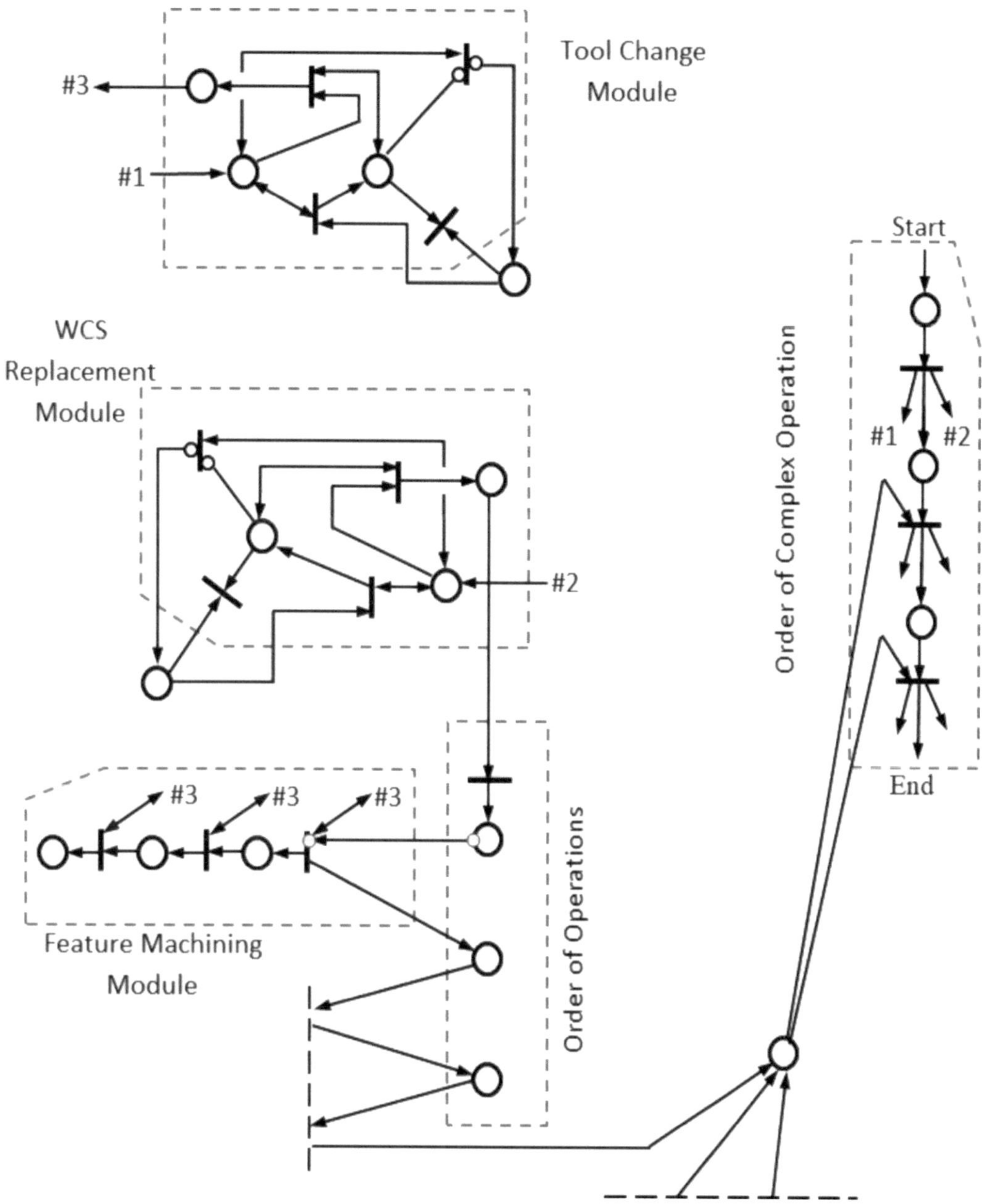

Figure 1. General (partial) multi-tool processing operation model.

The interdependence between M current status and M′ status directly resulting from it has been defined as follows:

$$M(p) = \begin{cases} 0 \; if \; p \in {}^{\cdot}t \; and \; p \notin t^{\cdot} \\ 1 \; if \; p \in t^{\cdot} \\ M(p) \; otherwise \end{cases} \tag{3}$$

where $t^{\cdot} - t$ transition output places set.

The aforementioned class is therefore a temporary Petri network with a binary marking function and inhibitor arcs. Inhibitor arcs make the model simpler, more transparent, therefore, easier to build and analyze. The suggested module structure allows for automation in generating the model. Modules for replacement of tools in the number of tools used for the process control the need to replace a tool. Topologically identical modules of control of calling the Workpiece Coordinate System (WCS) should be identified as the position required by the production process and the orientation of a processed object in a machine coordinate system. A module of complex operations is the core of the model. Each transition in the module calls a tool and WCS for the subsequent complex operation. A subsequent module is a sequence of operations conducted using the same tool on subsequent features. The modules is an area for structural optimization of the process by way of minimizing the number of tool changes. The model is supplemented by modules classifying simple operations for individual features. A marker on the last item of a module means ending processing of a given feature.

The aforementioned issue of optimizing the structure of multi-tool production operations might be solved using two simple heuristics: give preference to the tool that might currently process the largest possible number of areas and give preference to the tool that might currently complete all the operations that have been assigned to it. The suggested approach allows for a satisfying sequence of simple operations with a small number or iterations equal to the number of tools used. It is an optimal sequence in more than 95% cases. In order to appreciate the benefits of this method, one can compare it to the full browsing algorithm. The advantages of this method also include a very simple method of transferring its results in the form of an organized sequence of complex operations to a formal network model.

The sequence of processing individual features might be conditional on their technological nature. An opportunity to automatically classify the conducted operations then requires initiating the optimizing procedure. A function of priorities assigned to each transition is used for this purpose. Example use of this method based on genetic algorithm is shown in the work [22].

3. Variant process planning

Variant process planning consists of completing a certain scheme of indexing a designed part, which allows to find the most structurally and manufacturing similar component and take over its manufacturing process it with an insignificant number of adjustments. Such approach to process design shows its benefits when it is possible to classify manufactured machine parts in groups of parts, which are structurally and manufacturing similar. VCAPP is therefore closely linked to the group technology of machines and use of flexible manufacturing cells.

The effective use of flexible, robotized manufacturing cells is guaranteed by completion in such group technology idea systems. Hence, similarity of the machine parts processed there should be considered both at the stage of design, equipment and software configuration and during everyday use of the production cells. The Group Technology (GT) method has reached its mature form at the beginning of the 1990s [23]. Currently, GT is considered as a concept of production, in which production resource is functionally grouped in production cells for the purpose of processing machine parts with similar features in order to achieve a high level of production reproducibility and artificially extend the size of a production batch. GT is related to the cellular production [24] concept, which means production both in flexible manufacturing cells and in autonomous flex-cell. GT therefore involves a range of various operations of a design, production and organizational nature, which requires coherence and synchronization. Many of them, such as determining production similarity of machine parts, grouping and classification of machine parts, variant design, parametric programming of computer numerical control (CNC) machine tools, developing group manufacturing processes, designing group processing equipment, configuration of manufacturing cells [25] and the issue of planning and controlling the manufacturing process are still readily raised subject of development works.

Technical implementations of VCAPP systems most frequently include one of the two solutions: building a system based on finding a similar part of a previously manufactured machine part and adopting its production process plan or developing a group production process for the so-called synthetic representative of the group. In both cases, it is possible to use a process model developed in the Petri network technique; however, the second method is more natural in using its potential.

Feature precedence network (FPN) defined as a directed graph representing precedences that result from limitations imposed by the features is a precondition for designing a correct manufacturing operation. For complex parts, with a large number of processed features and relations between them, FPN might be very complex and very difficult to manually process. In [26, 27] has developed a system generating FPN based on the analysis of interactions between the features and verification of FPN using the Petri network model. The developed structure involves generating features by way of mapping from CAD system, analysis of interactions between the features and an algorithm automatically creating a Petri network corresponding to a given FPN. An analysis of interaction between the features involves comparing each pair of features with reference to the defined set of rules. These rules include heuristic preferences of processing sequence, which guarantee its effectiveness. The rules take geometric, production and economic factors into account.

In case of variant approach to production process design FPN represents a model developed for the synthetic representative of a given group of machines. The synthetic representative represents an abstract object that includes all types of features that can be found in the group of parts with production similarities. The process model should allow for an explicit definition of a processing task only by way of correct positioning of initial marking vector M_0. The sequence of processing individual features is defined based on the FPN. Subsequent, structurally unified modules of the model supplement the remaining functions of the model. The work [28] elaborates on this issue and gives an example of the variant approach for axially symmetrical parts.

4. Modeling and optimizing production systems

As it has been mentioned in Section 1, modeling, simulation and controlling the production process in automated, robotized, discrete production systems is the main area of using the Petri network technique in production. The systems are typical examples of asynchronous concurrent systems. The problem of management of limited resources, classifying tasks and autodiagnosis in such systems is a very complex issue and requires optimizing procedures both at the stage of design and during use. Building a model of a designed production systems allows to avoid a range of conceptual errors and facilitate modifications of a project. Simulation techniques implemented on abstract models allow to optimize the operation of a system, define its potential efficiency, identify any bottlenecks and define the actual utilization rate on production resource. Simulation of models developed in the Petri network technique allows to generate a network availability graph and, consequently, at the stage of using the system, to avoid downtime, diagnose the system and facilitate making correct decisions.

The following organized eight items are the appropriate class of Petri networks for the aforementioned tasks:

$$PN_2 = (P, T, E, W, I, S, R, M_0), \tag{4}$$

where P, T, E, I, S, R, M_0 components meet the PN_1, whereas W component is a function of multiplicity the arcs:

$W{:}(E \cup I) \rightarrow N; N{:}$set of natural numbers.

In this case, the conditions of preparing transition t for firing have the following form:

$$\begin{cases} M(p) > = W(p,t), \forall p \in \cdot t \\ M(p) < W(p,t), \forall (\mathrm{p},\mathrm{t}) \in \mathrm{I} \end{cases} \tag{5}$$

It is essential to remember that multiplicity of flow through the arc is assigned to both ordinary arcs assigned to E flow relation and to inhibitor arcs. Assigning multiplicity to inhibitor arcs—which is not a frequent case—allows for giving up another element: function of locations capacity. Inhibitor arcs might be used for controlling flow of markers, for instance, to avoid overflow of intra-operation buffers. Appropriate software of this class of Petri networks is an intuitive, flexible tool of modeling and simulation for an engineer. A lack of opportunity to generate random events, allowed by stochastic Petri networks is a certain limitation of the aforementioned Petri network class [18].

Software processing PN_2 class might be easily supplemented with a function optimizing the production system. The following example uses the genetic algorithm to find optimum priorities for individual transitions. The chromosome of each solution includes just as many genes as is found in the sum of transitions (**Figure 2**). Each of the genes is represented as a number between [0, 1], which corresponds to the priority value for the appropriate transition. The reproduction process for a new descendant includes random appropriate gene of their parents or, with certain low probability, the value of the gene is drawn from the range [0, 1], which corresponds to the mutation process. Details of the suggested algorithm have been put forward in the works [28, 29]. **Figure 2** shows an example of optimum chromosome for the production system presented in the further part.

The number of produced parts, the total number of order completion or, for example, the number of measured parts might be used as an optimization criterion. The last example refers to a situation, in which production efficiency is satisfying, there is a production

http://dx.doi.org/10.5772/intechopen.75135

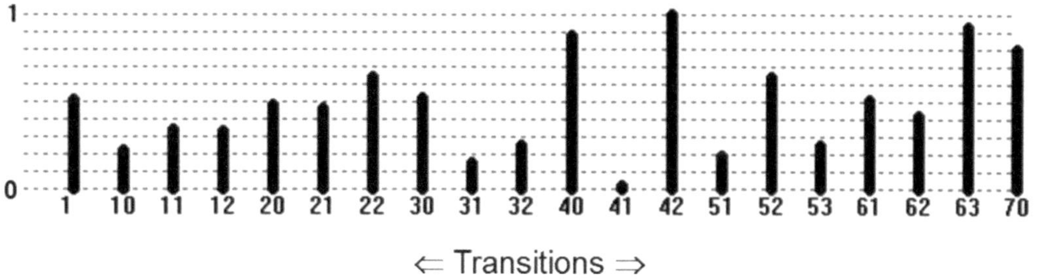

Figure 2. The best chromosome.

margin, so the manufacturer places more emphasis on improving production quality. It can be indirectly completed by attaching higher priority to transitions serving the measurement station.

The following example presents a production system model (**Figure 3**) with two lathes CNC (L_1, L_2), two drill-tap machining centers (M_1, M_2), two industrial robots (R_1, R_2), an input transporter (IN) and an output transporter (OUT), a measurement station (MS) and an intra-operation buffer. It has been designed for the purpose of processing rear axle guide pins—right (*B*) and left (*A*). *A* pin is processed at L_1 and M_1, *B* pin at L_2 and M_2. L_1 and L_2 lathes are identical, whereas M_1 processed four pins at the same time and M_2 only processes one.

Figure 4 presents Petri network for a robotized production system of processing rear axle guide pins. Marking the network corresponds to the production launch status. Only one transition (t_1) is possible in this status. The green spot means that it has at least one marker. $M_0(p_{30}) = 4$ should be set to ensure correct course of simulation. Arc multiplicity function has

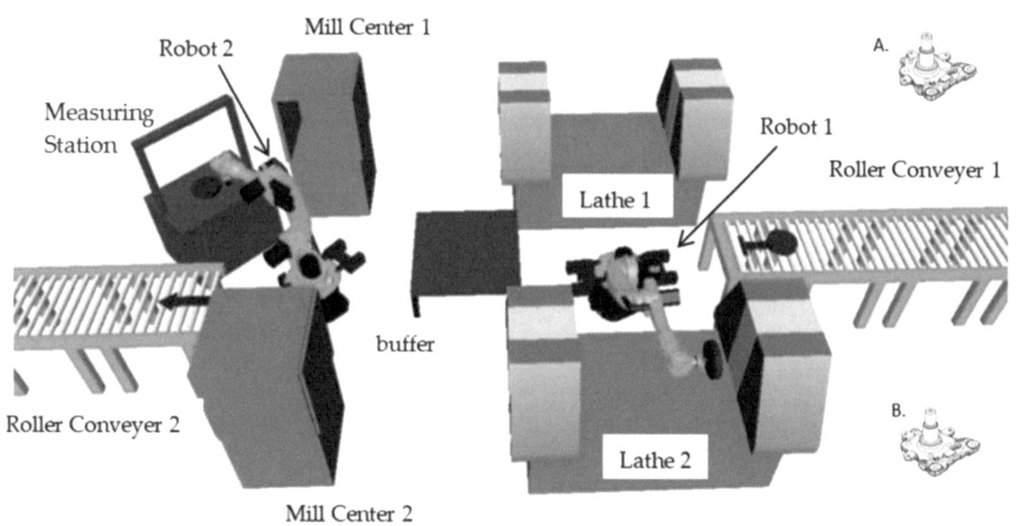

Figure 3. Robotized production system scheme.

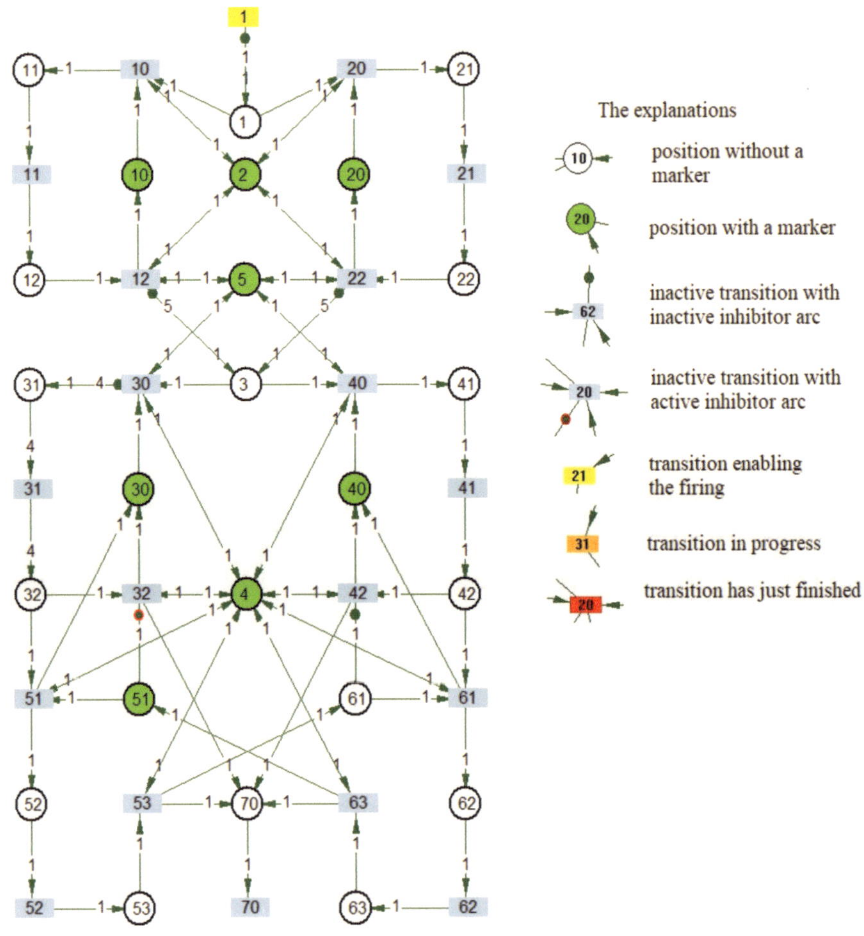

Figure 4. Robotized production system model.

been used for solving two problems: control of intra-operation buffer overflow and loading and unloading M_1. The buffer allows for storing up to five parts, therefore i(3, 12) and i(3, 22) inhibitor arcs have the multiplicity of 5, which results in blocking 12 and 22 transitions if p_3 has 5 markers. Position 5 aims at preventing simultaneous access to buffers of R_1 and R_2 robots. The i(31, 30) inhibitor arc, on the other hand, prevents loading non-processing parts to M_1 before all processed parts are unloaded. Multiplicity of 4th $p_{31}\rightarrow t_{31}$ arc allows to start processing only when M_1 includes a set of four parts. After completing t_{31} four markers are transferred to p_{32} and the unloading stage begins. Processed parts might be transferred to the input transporter (t_{32}, t_{42}) directly, or to the measurement station first, provided that it is expecting parts for measurement. The network topology ensures that parts processed at M_1 and M_2 are transferred for measurement. It is ensured by the marker flowing in the loop $p_{51}\rightarrow t_{51}\rightarrow p_{52}\rightarrow t_{52}\rightarrow p_{53}\rightarrow t_{53}\rightarrow p_{61}\rightarrow t_{61}\rightarrow p_{62}\rightarrow t_{62}\rightarrow p_{63}\rightarrow t_{63}\rightarrow p_{51}$.

Table 1 puts forward a description of module locations, whereas a description of transition of the presented model is put forward by **Figure 5**. The figure is an example of visualizing transformation results with a spreadsheet. Subchapter 6 puts forward the opportunities to integrate specialist software for Petri networks with other software packages.

http://dx.doi.org/10.5772/intechopen.75135

1	Next part is available	31	Part A before processing on M1
2	R1 is available	32	Part A after processing on M1
3	The buffer is not empty	40	M2 is available
4	R2 is available	41	Part B before processing on M2
5	Access to the buffer	42	Part B after processing on M2
10	L1 is available	51	A measurement request
11	Part A before processing on L1	52	A before measurement
12	Part A after processing on L1	53	A after measurement
20	L2 is available	61	B measurement request
21	Part B before processing on L2	62	B before measurement
22	Part B after processing on L1	63	B after measurement
30	M1 is available	70	Next part was made

Table 1. List of places from **Figure 4**.

T	text	tim	cycl	Kolum	Robot	Robot	Lathe1	Lathe2	MC1	MC2	MS
1	delivery of the next part	3	345	17.25							
10	loading L1	12	177	35.40	1						
11	machining on L1	120	176	352.00			1				
12	unloading L1	12	176	35.20	1						
20	loading L2	12	167	33.40	1						
21	machining on L2	120	167	334.00				1			
22	unloading L2	12	166	33.20	1						
30	loading M1	12	176	35.20		1					
31	machining on M1	540	43	387.00					1		
32	unloading M1	12	129	25.80		1					
40	loading M2	12	164	32.80		1					
41	machining on M2	150	163	407.50						1	
42	unloading M2	12	121	24.20		1					
51	transfer of A from M1 to MS	15	43	10.75		1					
52	measurement A	180	43	129.00							1
53	transfer of A from MS to OU1	15	42	10.50		1					
61	transfer of B from M2 to MS	15	42	10.50		1					
62	measurement B	180	42	126.00							1
63	transfer of B from MS to OU1	15	42	10.50		1					
70	transfer part	2	334	11.13							
				sum	137.2	160.25	352	334	387	407.5	255
				percentage	28.6	33.4	73.3	69.6	80.6	84.9	53.1

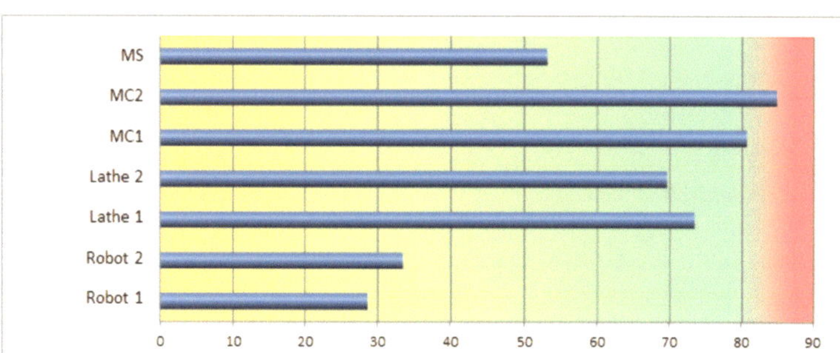

Figure 5. Spreadsheet of production resources utilization rate.

5. Petri networks as a manufacturing knowledge representation method

Contemporary CAPP systems have the architecture of expert systems [8]. Systems based on rules prevail among knowledge representation methods applied in expert systems. Petri networks do not constitute a turning point here. They are only another variant of the rule method. However, it is a ready carrier of knowledge with a large potential of expanding its capabilities, flexible as frames, able to capture the context to the extent not smaller than semantic networks, allowing for intelligent inference by way of introducing fuzzy rules. An advanced suggestion for using Petri networks for knowledge representation in CAPP systems has been put forward in the work [15], the authors of which present characteristics of the component of a rule approach to knowledge representation and an approach based on Petri network. Multilateral usefulness of Petri networks at various stages of building and using expert systems should be noted. Starting from a knowledge acquisition system [22, 30], to knowledge representation in a knowledge database [31, 32], user interface handling [33], inference mechanism [32, 34] to knowledge validation [33], the Petri network technique is helpful in all of these aspects.

Knowledge engineers most frequently use the structural formalism of knowledge representation based on the Petri network technique. Logical Petri Net (LPN) allows for easy modeling and verification of knowledge. The work [30] puts forward a fuzzy inference algorithm and a backward propagation algorithm for Adaptive Fuzzy Petri Network (AFPN). AFPN might be a finished platform for an expert system allowing for knowledge acquisition based on a teaching set.

The following example uses a maximally simplified version of AFPN, which still is a logical fuzzy Petri network (Eq.(6)) in the form of organized five units:

$$PT_3 = (P, T, E, \alpha, \mu), \tag{6}$$

where P is a set of facts: premises and conclusions; T: a set of rule cores; E: flow relation; α: $P \rightarrow [0, 1]$, association function which assigns a real value to each $p \in P$; and μ: $T \rightarrow [0, 1]$, certainty factor of the rule.

M_0 functions are replaced here with α component. A rule in PT_3 includes t transition and a set of premises (input locations) and conclusions (output locations). The fuzzy nature of inference results from using fuzzy aggregation functions used for calculating the certainty factor (CF) of the generated conclusions. If a conclusion is generated by only one rule (**Figure 6a**), we

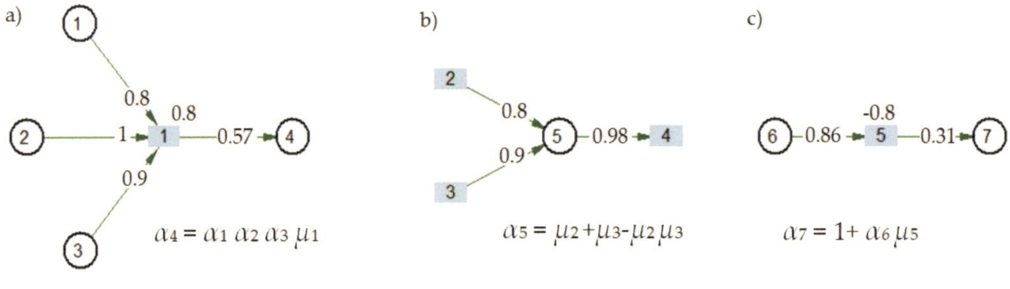

Figure 6. Setting the CF of conclusion.

http://dx.doi.org/10.5772/intechopen.75135

Place	status	descripion	CF
1	OUT	Use bar feeder	0.72
2	IN	High volume production	1.00
3	IN	Possible processing along the entire length	1.00
4	IN	Part is a shaft	1.00
5	OUT	Use centre lathe	0.25
6	IN	Part is short	0.86
7	OUT	Non-rigid part	0.31
8	OUT	Part is long	0.31
9	OUT	Use chucking lathe	0.77
10	OUT	Use steady rest	0.00
11	OUT	Use tailstock center	0.43
12	OUT	Use face drivers	0.23
13	OUT	Diameter is not constant	0.00
14	IN	Diameter is small	0.80
15	IN	Diameter is constant	1.00
16	OUT	Use follow rest	0.31

Table 2. List of places.

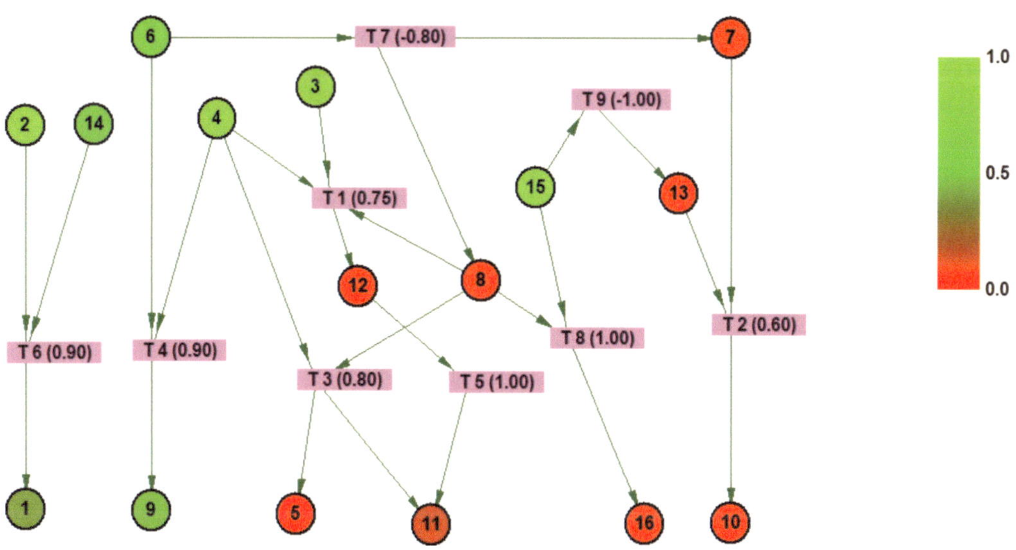

Figure 7. A fragment of knowledge database represented in the form of a petri network.

use fuzzy conjunction. Its CF is a product of premises of the rule and the CF of the very rule. If a conclusion is supported by two (**Figure 6b**) or more rules, fuzzy disjunction should be used. Its CF is a soft logical sum of indices generated by each of the rules separately. If the CF of transition is negative, what we have is a fuzzy negation (**Figure 6c**). All the three aggregation functions ensure maintaining CF between 0 and 1. Contrary to the previously presented PN_1 and PN_2 classes, PN_3 class cannot include a loop. It is also allowed that a graph is inconsistent. The threshold value of certainty indices is defined as global for the entire network. **Table 2** presents a description and list of CF for the shown example. **Figure 7** puts forward a fragment of the manufacturing knowledge database related to the method of mounting parts on a lathe, developed using the mechanism presented above.

6. Petri networks in computer integration manufacturing

Computer integration manufacturing (CIM) involves the integration of technical systems of constructing a product, planning a manufacturing process, technical preparation of production, programming CNC equipment, manufacturing, assembly, ensuring quality control, monitoring [35] and testing a product with facility and organizational systems of planning and controlling production in a company. Such integration should be implemented based on a shared, coherent system of databases: structural ones, ones for manufacturing processes, resource of production and transport, tools, processing and measuring equipment. Possible scope of modeling using the Petri network technique with respect to issues related to computer integration of manufacturing is very broad nowadays. The use of graphic tools such as the Update Petri Nets (UPN) class [36] based on Color Petri Networks (CPN) might be used as an example. UPN applications are systems based on rules, integrating control of information flow between CAD, CAPP, resources management and production flow.

A data exchange system between individual modules is a prerequisite for integration. It is still the weakest link of the contemporary computer integration systems. Currently, the bottleneck of integration might be removed, after introducing and popularization of the Automation Markup Language (AML). IEC 62714 [37] project available since 2014 is a solution for data exchange concentrating on the area of industrial automation. The document defines AML data exchange format based on Extensible Markup Language (XML) scheme data format. It has been developed to support data exchange for heterogeneous engineering tools in the environment of automated production resource. The objective of AML is to combine engineering tools of various fields, that is, mechanical engineering, workshop design, electric engineering, production process design, controlling course of production, development of human machine interface (HMI), programmable logic controller (PLC) programming, programming CNC machine tools and industrial robots, and so on. AML might describe both physical and logical aspects of all components of a production system, their topology, geometry, kinematics, logics of control, and represent the links and hierarchy of all objects considered. It should be emphasized that XML format serving as a base for AML language has been developed since 1990, mostly as a spontaneous initiative of engineers dissatisfied with previously offered solutions, including standard for the exchange of product model data (STEP). The success of XML format is a result of flexibility of the language, ease of defining complex data structures, lack of barriers in other suggested solutions, its alphanumeric character, ease of control and direct analysis, open and intuitive character. Solutions have been put forward for several years, also in the form of standards, presenting data representation for the Petri network model in a format based on the Petri Net Markup Language (PNML) [38]. PNML has been introduced as a data exchange format for all Petri network classes. All of this is a fundamental premise for formulating a favorable forecast for acceptance and quick popularization of AML language as a data exchange language in industrial automation and its environment.

Figure 8 shows a scheme of an integrated production unit design system, taking into account data exchange streams and main functions of individual modules. Attention should be drawn

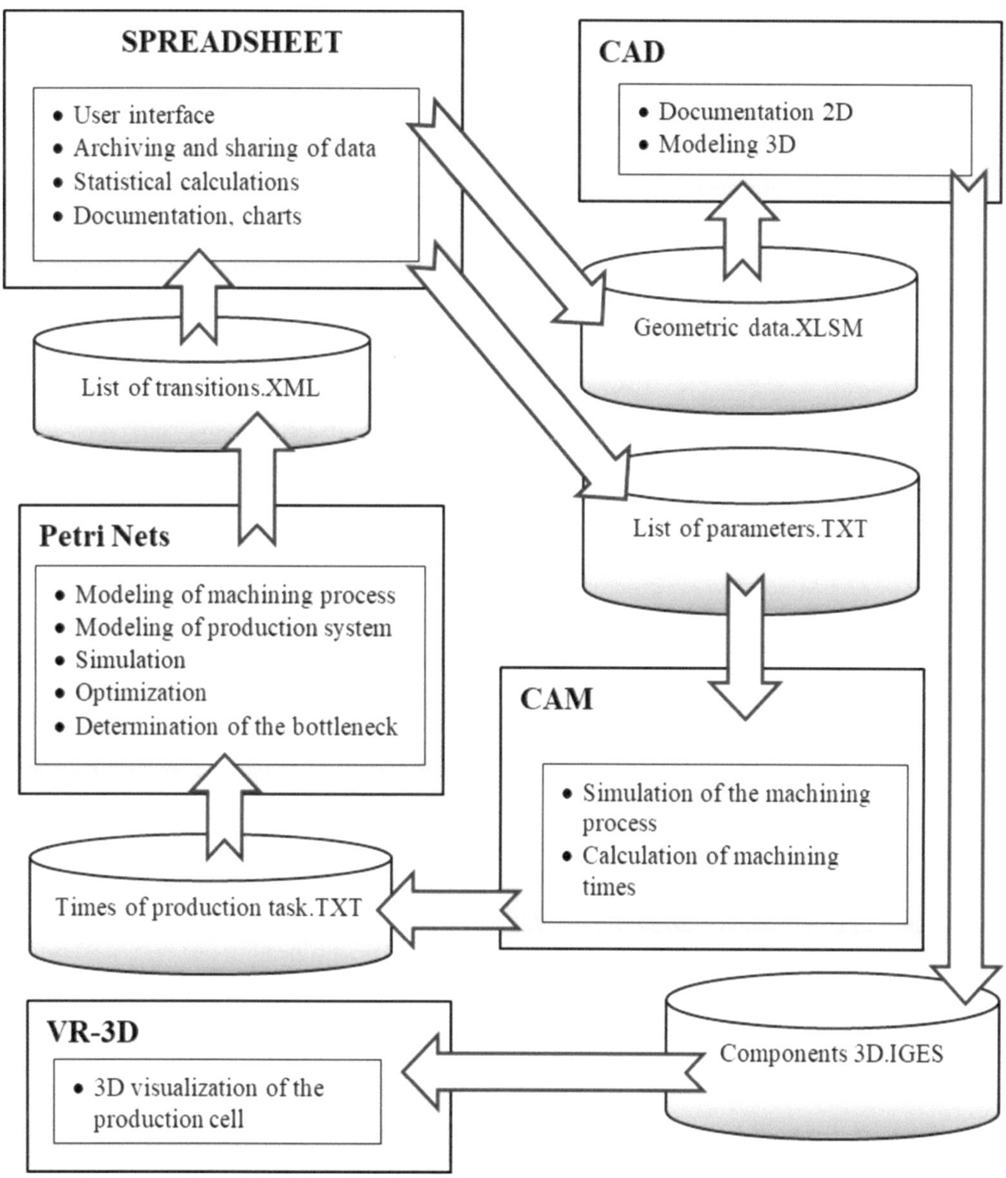

Figure 8. Data exchange streams in integrated manufacturing cell planning system.

to the role of a spreadsheet with active macro application option, which allows to compile input data for computer aided design (CAD) and computer aided manufacturing (CAM) systems and control their correctness, completeness, consistency and adherence with the allowed range. Diagnostics of data is a significant element that influences comfort of using the design system. Full integration of a spreadsheet with popular CAD systems is possible nowadays, which significantly facilitates structural designing and drawing up documentation in the form of drawings. At the same time, a spreadsheet allows for drawing up a data file for a parametrized program of processing a selected class of produced machine parts by using related data sheets. An Excel Macro-Enable Workbook (XLSM) format file allows to use macros, that is, command sequences that might be used for the automation of completed functions in a spreadsheet. A 3D model is made by preparing a drawing in CAD program, which is integrally linked to the data included in the spreadsheet's data file. Then, 2D documentation is made based on such model, if it is necessary. The code generated in a spreadsheet has a simple structure in the form of a list of R-parameters plus the name of a processed part. These are input data for CAM system and/or the numerical control system of the machine tool, where parametric programs related to the aforementioned list of variable parameters are archived. Aside from formal verification of the controlling program's form, processing in CAM aims at providing a simulation of processing. Aside from confirming correctness of programmed trajectories of a machine's movement, the simulation made provides an exact completion time for individual processing operations. The results are exported to the Petri Nets Modeler (PNM) package, which allows for exact time simulation of a production unit's work. A list of Petri network model transitions in XML format are again transferred to the spreadsheet, thus closing the process chain. It includes a list of transitions with their completion times and a number of repetitions in a given period. A spreadsheet allows to easily obtain a clearly structured document presenting, for example, utilization rate on production resource (**Figure 5**).

PNM package is a key module of the presented design system. Its objective consists of computer simulation, modeling and optimizing of both individual and group processes of production, dedicated to autonomous flex-cells, multi-machine tool stations and robotized manufacturing systems.

7. Summary

Both the very process of planning manufacturing processes and the structure of the manufacturing processes require applying a conditional-event mechanism that is naturally represented in the Petri networks technique. This chapter briefly outlines the broad opportunities of applying this technique to design of discrete production processes. The period 1990s are the most intense development period of Petri networks technology. We are currently at the stage of applying already mature programming tools, increasingly frequently integrated in heterogeneous planning systems. The greatest advantages of Petri networks technique visible and acknowledged by engineers include: an opportunity to generate models with various levels of detail, graphic user interface at the stage of acquisition, verification and processing a model, ease of modification and expansion of a model using its hierarchical and block character. The fact that each model contains production knowledge is the greatest advantage of applying modeling techniques in the planning of production processes. Hence, the premises for automation of the design process, at least in the area of implementing routine works. An engineer's intellectual efforts might therefore be aimed at introducing innovative solutions and improvements.

Author details

Roman Stryczek

Address all correspondence to: rstryczek@ath.bielsko.pl

University of Bielsko-Biala, Poland

References

[1] Ham I, Lu CY. Computer-aided process planning: The present and the future. Annals of the CIRP. 1988;**37**:1-11. DOI: 10.1016/S0007-8506(07)60756-2

[2] Cecile JA, Srihari K, Emerson CR. A review of petri net application in process planning. International Journal of Advanced Manufacturing Technology. 1992;**7**:168-177. DOI: 10.1007/BF02601620

[3] Kiritsis D. Petri net modelling for dynamic process planning. In: Vernadat F, editor. Integrated Manufacturing Systems Engineering. 1995. pp. 206-218. DOI: 10.1007/978-0-387-34919-0_14

[4] Kiritsis D. A review of knowledge-based expert systems for process planning. Methods and problems. International Journal of Advanced Manufacturing Technology. 1995;**10**:240-262. DOI: 10.1007/BF01186876

[5] Kiritsis D, Porchet M. A generic petri net model for dynamic process planning and sequence optimization. Advances in Engineering Software. 1996;**25**:61-71. DOI: 10.1016/0965-9978(95)00086-0

[6] Kiritsis D, Xirouchakis P, Gunther C. Petri net representation for the process specification language. In: Proceedings of International Workshop on Intelligent Manufacturing Systems (IMS-EUROPE); 1998; Lausanne. Available from: https://www.researchgate.net/publication/268259692 [Accessed: 2015-01-19]

[7] Kiritsis D, Neuendorf KP, Xirouchakis P. Petri net techniques for process planning cost estimation. Advanced in Engineering Software. 1999;**30**(6):375-387. DOI: 1016/S0965-9978(98)00126-4

[8] Lee DH, Kiritsis D, Xirouchakis P. Iterative approach to operation selection and sequencing in process planning. International Journal of Production Research. 2004;**42**:4745-4766. DOI: 10.10180/00207540410001720412

[9] Xirouchakis P, Kiritsis D, Persson JG. A Petri net technique for process planning cost estimation. Annals of the CIRP. 1998;**47**:427-430. DOI: 10.1016/S0007-8506(07)62867-4

[10] Xirouchakis P, Kiritsis D, Gunther C, Persson JG. A petri net technique for batch delivery time estimation. CIRP Annals—Manufacturing Technology. 1999;**48**:361-364. DOI: 10.1016/S0007-8506(07)63202-8

[11] Horvát L, Rudas IJ. Manufacturing process planning using Object Oriented Petri nets supported by entropy based fuzzy reasoning. In: Proceedings of IFAC Conference; 27-29 September 1994; Baden-Baden: Integrated Systems Engineering; 1995. pp. 311-316. DOI: 10.1016/B978-0-08-042361-6.50054-4

[12] Horvát L, Rudas IJ, Camarinha-Matos LM. Structured model representation of manufacturing process using petri nets and knowledge based tools. In: Balance Automation Systems II. Boston: Springer; 1996. pp. 472-480. DOI: 10.1007/978-0-387-35065-3_47

[13] Rudas IJ, Horvát L. Modelling of manufacturing processes using a petri-net representation. Engineering Applications of Artificial Intelligence. 1997;**10**(3):243-255. DOI: 10.1016/S0952-1976(97)00006-7

[14] Horvát L, Rudas IJ. Evaluation of Petri net process model representation as a tool of virtual manufacturing. In: Proceedings of the Conference: Systems, Man, and Cybernetics; November 1998; IEEE; 1998. DOI: 10.1109/ICSMC.1998.725405

[15] Lee KH, Jung MY. Flexible process sequencing using petri net theory. Computers & Industrial Engineering. 1995;**28**(2):279-290. DOI: 10.1016/0360-8352(94)00191-O

[16] Lee DH, Kiritsis D, Xirouchakis P. Search heuristics for operation sequencing in process planning. International Journal of Production Research. 2001;**39**:3771-3788. DOI: 10.1080/00207540110061922

[17] Tuncel G, Bayhan GM. Applications of petri nets in production scheduling: A review. International Journal of Advanced Manufacturing Technology. 2007;**34**:762-773. DOI: 10.1007/s00170-006-0640-1

[18] Al-Ahmari A, Li Z. Analysis of multimachine flexible manufacturing cell using stochastic petri nets. Advances in Mechanical Engineering. 2016;**8**(11):1-9. DOI: 10.1177/1687814016680168

[19] Uzam M. The use of the petri net reduction approach for an optimal deadlock prevention policy for flexible manufacturing systems. International Journal of Advanced Manufacturing Technology. 2004;**23**:204-219. DOI: 10.1007/s00170-002-1526-5

[20] Kumar RR, Singh A, Tiwari MK. A fuzzy based algorithm to solve the machine-loading problems of a FMS and its neuro fuzzy petri net model. International Journal of Advanced Manufacturing Technology. 2004;**23**:318-341. DOI: 10.1007/s00170-002-1499-4

[21] Ahson SI. Petri net models of fuzzy neural networks. IEEE Transaction Systems Manufacturing and Cybernetic. 1995;**25**(6):926-932. DOI: 10.1109/21.384255

[22] Stryczek RA. Hybrid approach for manufacturability analysis. Advances in Manufacturing Science and Technology. 2011;**35**(3):55-70. Available from: http://advancesmst.prz.edu.pl/

[23] Kusiak A. Group Technology. Intelligent Design and Manufacturing. New York: Wiley; 1992

[24] Suresh NC, Kay JM, editors. Group Technology and Cellular Manufacturing: A State-Of the-Art Synthesis of Research and Practice. USA: Springer; 1998. DOI: 10./1007/978-1-4615-5467-7

[25] Kou Y, Yang J. Two-stage optimization of manufacturing cell. In: Proceedings of the First International Conference on Advances in Swarm Intelligence; 12-15 June 2010; Beijing: ICCPR; 2010. pp. 281-289. DOI: 10.1007/978-3-642-13495-1_35

[26] Sormaz DN, Khoshnevis B. Modeling of manufacturing feature interaction for automated process planning. Journal of Manufacturing Systems. 2000;**19**(1):28-45. DOI: 10.1016/S0278-6125(00)88888-3

[27] Sormaz DN, Thiruppalli S, Arumugam J. Evaluation of manufacturing feature precedence constraints using Petri nets. In: Proceedings of the 11th Industrial Engineering Research Conference; Orlando. 2002. Available from: www.researchgate.net/publication/2534056

[28] Stryczek R. Petri nets in computer aided group technology. In: Zawiślak S, Rysiński J, editors. Graph-Based Modelling in Engineering. Mechanisms and Machine Science 42. Springer; 2017. pp. 143-164. DOI: 10.1007/978-3-319-39020-8_11

[29] Stryczek R. A meta-heuristics for manufacturing systems optimization. Advances in Manufacturing Science and Technology. 2009;**33**(2):23-32. Available from: http://advancesmst.prz.edu.pl/

[30] Li X, Yu W, Lara-Rosano F. Dynamic knowledge interface and learning under adaptive fuzzy petri net framework. IEEE Transactions on Systems Man and Cybernetics. 2000;**30**(4):442-450. DOI: 10.1109/5326.897071

[31] Garg ML, Ashon SI, Gupta PV. A fuzzy petri net for knowledge representation and reasoning. Information Processing Letters. 1991;**39**:165-171. DOI: 10.1016/0020-0190(91)90114-W

[32] Liu HC, You JX, Li ZW, Tian G. Fuzzy petri nets for knowledge representation and reasoning: A literature review. Engineering Applications of Artificial Intelligence. 2017;**60**:45-56. DOI: 10.1016/j.engappai.2017.01.012

[33] Stryczek R. Petri net-based knowledge acquisition framework for CAPP. Advances in Manufacturing Science and Technology. 2008;**32**(2):21-38. Available from: http://advancesmst.prz.edu.pl/

[34] Yang SJH, Chu WC, Lee J, Huang WT. A fuzzy Petri net mechanism for fuzzy rules reasoning. In: Proceedings of the 21st International Computer Software and Application Conference (COMPSAC'97); 11-15 August 1997; Washington: IEEE; 1997. pp. 438-443. DOI: 10.1109/CMPSAC.1997.625031

[35] Miyagi PE, Riascos LAM. Modeling and analysis of fault-tolerant systems for machining operations based on petri nets. Control Engineering Practice. 2006;**14**:397-408. DOI: 10.1016/S1474-6670(17)32785-4

[36] Harhalakis G, Lin CP, Mark L, Muro-Medrano PR. Structured representation of rule-based specifications in CIM using updated petri nets. IEEE Transactions on System Manufacturing and Cybernetics. 1995;**25**(1):130-144. DOI: 10.1109/21.362959

[37] IEC 62714-1:2014. Engineering data exchange format for use in industrial automation systems engineering—Automation Markup Language. International Electronical Commission. Available from: http://reference.globalspec.com/standard/3889214/iec-62714-1-2014

[38] Hillah LM, Kindler E, Kordon F, Petrucci L, Trèves N. A primer on the Petri net markup language and ISO/IEC 15909-2. In: Proceedings of the CPN Workshop; October 2009; Aarhus. Available from: http://www.pnml.org/papers/pnnl76.pdf

Chapter 4

Reliability Evaluation for Mechanical Systems by Petri Nets

Jianing Wu and Shaoze Yan

Additional information is available at the end of the chapter

http://dx.doi.org/10.5772/intechopen.79624

Abstract

The current trend in mechanical engineering is to design mechanical systems with higher stability, reliability, availability and operability. In order to meet the requirement of high reliability for a machine, it is of great importance for designers to seek the weak links in the system and learn the state of the key subsystems, carrying out the remedial measures when necessary. Hence, behavior modeling and failure analysis are the two aspects seriously concerned in the reliability evaluation in mechanical systems. This chapter will introduce new methodologies that use the fuzzy reasoning Petri net (FRPN) models to evaluate the reliability of mechanical systems in reliability prediction, reliability apportionment and reliability analysis. Cases are proposed by analyzing a spacecraft solar array system using the proposed method. Results indicate that the Petri nets models can contribute to a higher accuracy in reliability evaluation for mechanical systems.

Keywords: reliability evaluation, mechanical system, Petri nets

1. Introduction

Some mechanical systems experience complicated environment which may continuously influence the reliability and availability. For instance, the spacecraft solar arrays are one of the most vital links to satellite mission success because providing reliable power over the anticipated mission life is critical to all satellites [1–3]. Although the faults have been reduced in the last few years by some measures, it still affects the longevity of the satellite severely, and faults of mechanical system occupy a large proportion of all the anomalies [3]. As a result, it is necessary for mechanical systems to evaluate reliability in different stages, including conceptual design of mechanical system, initial design and system improvement. The tasks of reliability evaluation in these stages are defined as reliability prediction, reliability apportionment and reliability analysis, respectively. Many methodologies such as reliability block diagram (RBD), failure mode effect analysis (FMEA) and fault tree analysis (FTA) are widely used in reliability evaluation for electronic systems [4–6]. Recently, a number of papers reported the methodologies that use these models to evaluate reliability of the mechanical systems [7–9]. However, there still has some obstacles needed to be overcome for reliability evaluation of mechanical systems. Generally, three tasks should be accomplished, including reliability prediction, reliability apportionment and reliability analysis. We summarize the defects of previous research from the three aspects mentioned above.

For reliability prediction, there are currently four main ways of reliability prediction for mechanical systems [10–12], including the similar product method (SPM), correction coefficient method (CCM), analysis of physics reliability method (APR) and parts count reliability prediction (PCRP). However, in the phase of conceptual design stage for one complex mechanical system, there has no enough experimental data or field record because the machine is not physically built. Moreover, APR is based on the physical failure mechanism which cannot be clearly identified in the conceptual design stage.

For reliability apportionment, there are two important issues needed to be addressed, i.e. how to describe the relationship among the different components and how to overcome data deficiency problem in the early stage of design [13–16]. It is usually hard to describe the factors of one mechanical system by the binary logic because the state cannot be simply classified into function or failure. Further, since the lack of system reliability data is a commonly encountered case in the initial stage of design, the reliability apportionment merely based on mathematics may not be feasible.

For reliability analysis, the FTA model has been widely employed as a powerful technique to evaluate the safety and reliability of complex systems by many scholars [17–19]. However, FTA has some limitations in reliability analysis. Firstly, in FTA, the probabilities of basic events must be known before analysis, but the designers can hardly obtain the probability of each fault because the conventional reliability test of the solar array mechanical system is difficult to carry out [19]. Secondly, it is not easy for FTA to conduct further quantitative analysis automatically due to the lack of effective means of mathematical expression. Thirdly, FTA cannot find the weak links of the system precisely, describe the propagation of fault and represent the characteristics of the system before and after improvement. In the literature, fuzzy reasoning is an effective method to solve the above problems [20].

The Petri net is one of the mathematical modeling approaches for the description of distributed systems, which consists of places, transitions, and directed arcs [21–23]. Many extensions to the Petri nets have been successfully applied in analyzing reliability of mechanical systems [24]. The fuzzy reasoning Petri net is a mathematical and graphical combined tool that can build a complex system with a variety of logical connections by using fuzzy reasoning, which may fit for building the reliability model for mechanical systems and evaluating reliability of them [20]. As a result, the primary objective of this chapter is to introduce the FRPN based models to evaluate the reliability of mechanical systems, including reliability prediction, reliability apportionment and reliability analysis. Some cases are included to illustrate the effectiveness of the methods.

2. Fuzzy reasoning Petri net

A great volume of literature combines fuzzy reasoning and Petri net to accomplish the fault diagnosis and reliability analysis [25–27]. Gao presented an FRPN model and proposed a fuzzy reasoning algorithm based on matrix equation expression [19]. An FRPN model can be defined as an 8-tuple model instead of the basic 5-tuple Petri net model [19].

1. Places, namely, a set of propositions,

$$P = \{p_1, p_2 \cdots p_n\}, 1 \times n; \tag{1}$$

2. Transitions,

$$R = \{r_1, r_2 \ldots r_m\}, 1 \times m; \tag{2}$$

3. Directed arcs propositions to rules,

$$I : P \times R \to \{0, 1\}, n \times m \tag{3}$$

4. Directed arcs from rules to propositions,

$$O : P \times R \to \{0, 1\}, n \times m \tag{4}$$

5. Complementary arcs from positions to rules,

$$H : P \times R \to \{0, 1\}, n \times m \tag{5}$$

6. Truth degree vector:

$$\theta = (\theta_1, \theta_2, \cdots \theta_n)^T, \theta_i \in [0, 1], n \times 1 \tag{6}$$

7. Marking vector:

$$\gamma : P \to \{0, 1\}, \gamma = (\gamma_1, \gamma_2 \cdots \gamma_n)^T, n \times 1 \tag{7}$$

8. Confidence of

$$r_j : C = diag\{c_1, c_2 \ldots c_{25}\}, 1 \times m. \tag{8}$$

On the basis of algorithm provided by Gao [19], the simulation can be operated automatically. The following are the main rules:

1. If one transition is fired, the token will be sent to the upper place.
2. If there are many places to one transition like AND gate in FTA model, the upper truth value will be the minimum; if there are many places to many transitions like OR gate in FTA model, the upper truth value will be the maximum.
3. The vector $\gamma = (\gamma_1, \gamma_2 \cdots \gamma_i \cdots \gamma_n)^T, n \times 1$ shows the propagation of the faults in model. If the element $\gamma_i = 1$, the place p_i will get the token.
4. The truth degree vector $\theta = (\theta_1, \theta_2, \cdots \theta_n)^T$ shows the fuzzy possibility of the faults.

The PRPN model takes advantage of the following maximum algebra

1. $\oplus : A \oplus B = D$, where A, B and D are all $m \times n$ dimensional matrices, such that

$$d_{ij} = \max\{a_{ij}, b_{ij}\} \tag{9}$$

2. $\otimes : A \otimes B = D$, where A, B and D are $m \times p$, $p \times n$ and $m \times n$-dimensional matrices respectively, such that

$$d_{ij} = \max_{1 \le k \le p} \{a_{ik} \cdot b_{kj}\} \tag{10}$$

The firing and control vectors are stated as follows [19]:

$$\begin{cases} \mu_{m\times 1}^k = 1_{m\times 1} - (I+H)^T \otimes \overline{\gamma}^k \\ \rho_{m\times 1}^k = 1_{m\times 1} - \left(I^T \otimes \left(\overline{\gamma}^k \oplus \overline{\theta}^k\right)\right) \oplus \left(H^T \otimes \left(\overline{\gamma}^k \oplus \theta^k\right)\right) \end{cases} \tag{11}$$

in which

$$\begin{cases} \overline{\theta}^k = 1_{\mathrm{m}\times 1} - \theta^k \\ \overline{\gamma}^k = 1_{\mathrm{m}\times 1} - \gamma^k \end{cases} \tag{12}$$

The marking and truth degree vectors can be obtained by

$$\begin{cases} \gamma^{k+1} = \gamma^k \oplus [O \otimes \mu] \\ \theta^{k+1} = \theta^k \oplus [(O \cdot C) \otimes \rho] \end{cases} \tag{13}$$

which reflects the status of the components in the mechanical system. The FRPN model is suitable to describe the status transition in a mechanical system because

1. The FRPN model is constructed by the places and logical connections which match the properties of mechanical systems with multiple components.

2. The FRPN model can describe the fault propagation in mechanical system by fuzzy reasoning, which can describe the properties of mechanical systems accurately.

3. The FRPN model is based on an iteration algorithm, so the status transition can be easily tracked, which may be useful for examining the fault propagation and fault severity in the system.

3. Reliability evaluation by FRPN

For evaluating the reliability of a mechanical system, one should complete a series of work including reliability prediction in the stage of conceptual design, reliability apportionment in the stage of initial design, and reliability analysis in the stage of system improvement. The following subsections will illustrate the method of how to evaluate reliability by FRPN models.

3.1. Reliability prediction by FRPN

3.1.1. Method

Reliability prediction acts when a product is in the stage of conceptual design. Here we introduce a method of reliability prediction of mechanical systems. This method includes the following steps (**Figure 1**). First, we will build an FRPN model of the mechanical system by its working principle and the logical connections among the components. Second, we get three key values which characterize quantity, importance and quality of the components in the mechanical system. Third, we will arrive at the reliability prediction result by parts count reliability prediction (PCRP). Finally, the reliability prediction formula of mechanical system denotes to

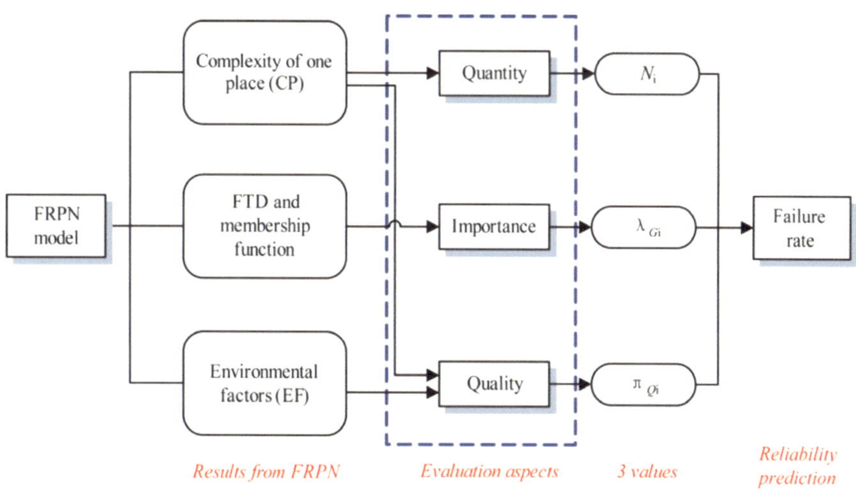

Figure 1. Main process of reliability prediction.

$$\lambda_p = \sum_{i=1}^{T} N_i \cdot \lambda_{\mathrm{Gi}} \pi_{\mathrm{Qi}} \tag{14}$$

where λ_p is the final predicted failure rate, λ_{Gi} and π_{Qi} are the indexes which indicate importance and quality of the components [28].

3.1.2. Case study

We take the deployable solar array used in spacecraft as an example. The running process of a typical deployable solar array is shown in **Figure 2**, which is widely used for power supply in the spacecraft nowadays. In general, the entire running process includes three stages, i.e. the deployable solar array is first folded, then deployed in the orbit and finally oriented to the sun to generate power for satellite.

In general, the mechanical system of the solar array consists of seven kinds of mechanisms [29–31], i.e. the hold-down and release mechanism, the solar panel, the driving mechanism, the deployable mechanism, the locking mechanism, the synchronization mechanism, and the orientation mechanism, as shown in **Figure 3**. Torsion spring is often chosen to drive the solar array, the closed cable loop (CCL) is used as the synchronization mechanism, and the stepping motor or servo motor is carried to orient to the sun. The driving mechanism, the deployable mechanism and the locking mechanism are always integrated into the hinge. Therefore the five main mechanisms of the solar array include hold-down and release mechanism, the solar panel, the hinge, the synchronization mechanism and the orientation mechanism.

We use R_1 to R_5 to represent the reliability of the five mechanisms, respectively. Then the reliability of the mechanical system can be calculated as follows:

$$R = R_1 R_2 R_3 R_4 R_5 \tag{15}$$

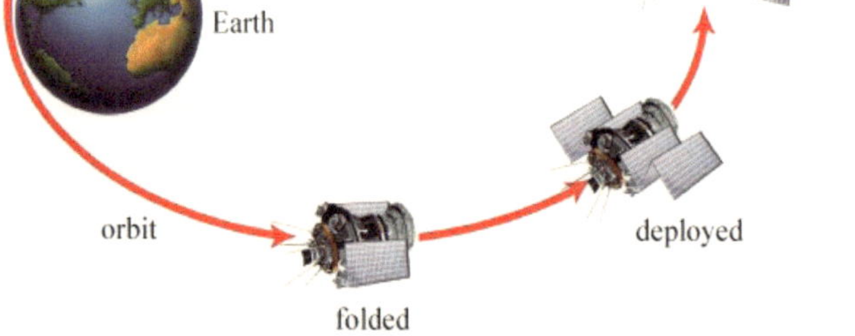

Figure 2. Operating principle of a deployable solar array.

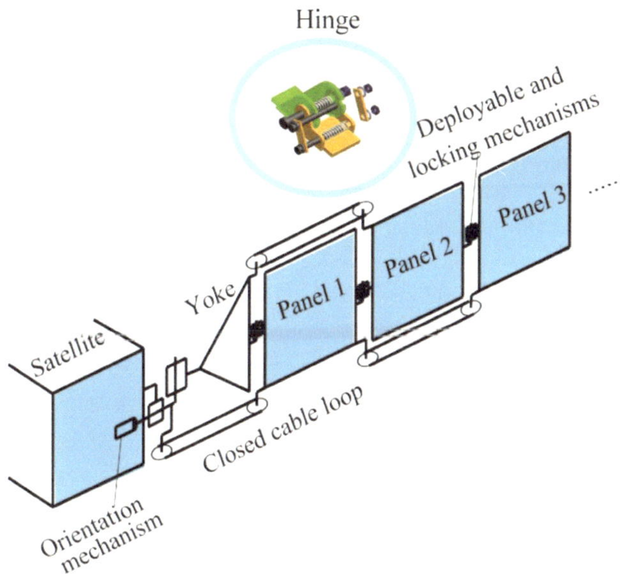

Figure 3. Mechanisms in a spacecraft solar array.

In the phase of conceptual design, designers should divide the reliability of the system into the five main parts. The following section introduces a new method of reliability apportionment which focuses on how to get the predicted values of $R_i(i = 1, 2, 3, 4, 5)$ to meet the requirement of the design standard. We build an FRPN model for the mechanical system of the solar array (**Figure 4**). **Table 1** shows the markers and events of FRPN model [32, 33].

By the method shown in **Figure 1**, we can measure the complexity of the *i*th place (CP) as a number of N_i, the final truth degree of the *i*th place (FTD) as λ_{Gi}, and the environmental factor

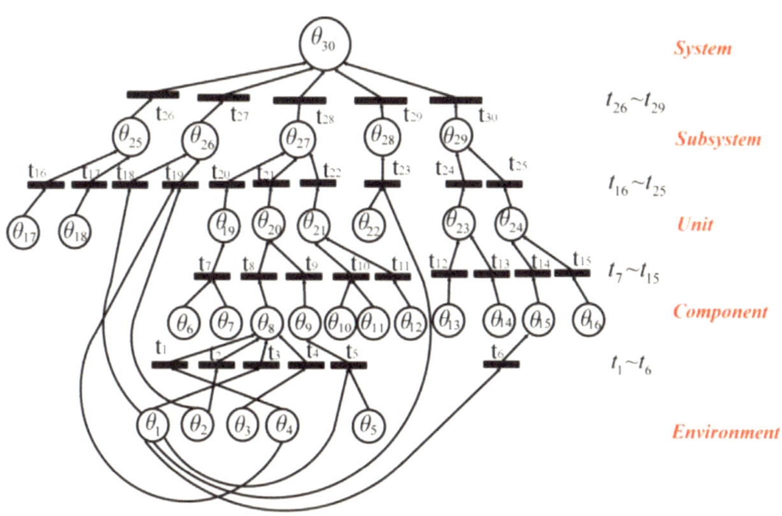

Figure 4. The FRPN model of the solar array for reliability prediction.

Marker	Event	Truth degree	Marker	Event	Truth degree
P_1	Harsh thermal environment in space	0.9	P_{16}	Fault of the bearing in the reducer	0.4
P_2	Vacuum and micro-gravity environment in space	0.6	P_{17}	Fault of the electronic arcing of the hold-down and release mechanism	0.7
P_3	Fault of the grease used in hinges between panels	0.4	P_{18}	Fault of the cutter of the hold-down and release mechanism	0.7
P_4	Impact caused by particles in space	0.7	P_{19}	Fault of the driving mechanism	–
P_5	Fault of the brass gasket	0.5	P_{20}	Fault of the deployable mechanism	–
P_6	Fault of the main driving torsion spring	0.6	P_{21}	Fault of the locking mechanism	–
P_7	Fault of the reserved driving torsion spring	0.6	P_{22}	Fault of the steel wire	0.7
P_8	Fault of the driving pin in the hinge	–	P_{23}	Fault of the stepping motor	–
P_9	Fault of the side wall of the hinge	–	P_{24}	Fault of the transmission system	–
P_{10}	Fault of the main locking spring	0.8	P_{25}	Fault of the hold-down and release mechanism	–
P_{11}	Fault of the reserved locking spring	0.5	P_{26}	Fault of the solar panels	–
P_{12}	Fault of the locking pin of the hinge	0.5	P_{27}	Fault of the hinges	–
P_{13}	Fault in the mechanical part of the stepping motor	0.3	P_{28}	Fault of the synchronization mechanism	–
P_{14}	Fault in the electronic part of the stepping motor	0.2	P_{29}	Fault of the orientation mechanism	–
P_{15}	Fault of the gear in the reducer	–	P_{30}	Fault of the mechanical system of the solar array	–

Table 1. Markers and events of FRPN model for reliability prediction.

(EF) of the *i*th place as π_{Qi}. Some details can be checked in [32]. We collected the actual reliability data (lifetime of mechanical systems) of the solar arrays in a group of satellites from 1950s to 2000s provided by [34]. The results show that all of the predicted reliability lies in the interval of the operation data, which demonstrates the correctness of FRPN-based model for reliability prediction. **Figure 5** validates the predicted reliability by using the four selected time: 0.025×10^6 h, 0.05×10^6 h, 0.075×10^6 h and 0.1×10^6 h. Some more details can be checked in [34].

3.2. Reliability apportionment by FRPN

3.2.1. Method

After reliability prediction in the conceptual design phase, the engineer should start reliability apportionment that acts when a product is in the stage of initial design. The conventional

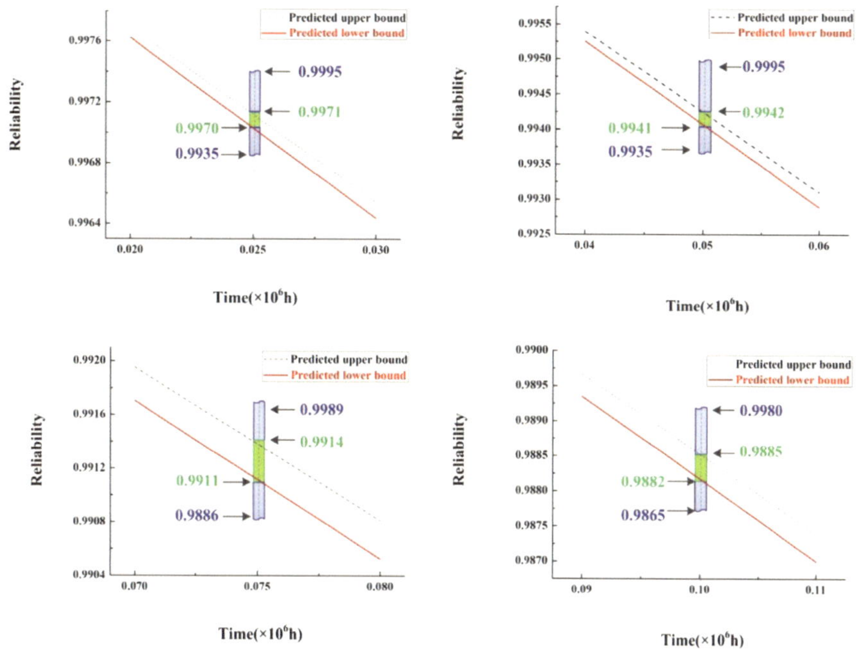

Figure 5. Comparison between the predicted reliability and real reliability at selected phases.

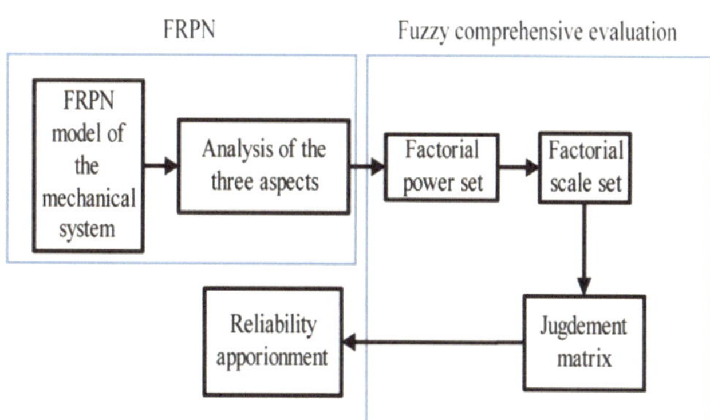

Figure 6. Procedures for reliability apportionment by FRPN. The FRPN model is used in the first and second steps and the following two steps use the fuzzy comprehensive evaluation.

reliability apportionment approaches including equal distribution method, Alins distribution method and algebra distribution method are widely used in the early stage of the reliability design [35, 36]. However, these methods have some limitations. It is obvious that dividing the system reliability into those of the subsystems equally may ignore the diversity of the components. Although the Alins distribution method and the algebra distribution method involve the importance or complexity of the different units, they are heavily dependent on the existing data and engineering experience which are scare in the early stage of the reliability design. Here we propose an FRPN-based method for reliability apportionment to solve the problems discussed above. This method includes the following steps (**Figure 6**):

1. Decompose the mechanical system;
2. Build the FRPN model of the mechanical system;
3. Analyze the three aspects including the complexity of one component during propagation of the faults, the importance of one component and the working environment;
4. Fuzzy comprehensive evaluation;
5. Reliability apportionment.

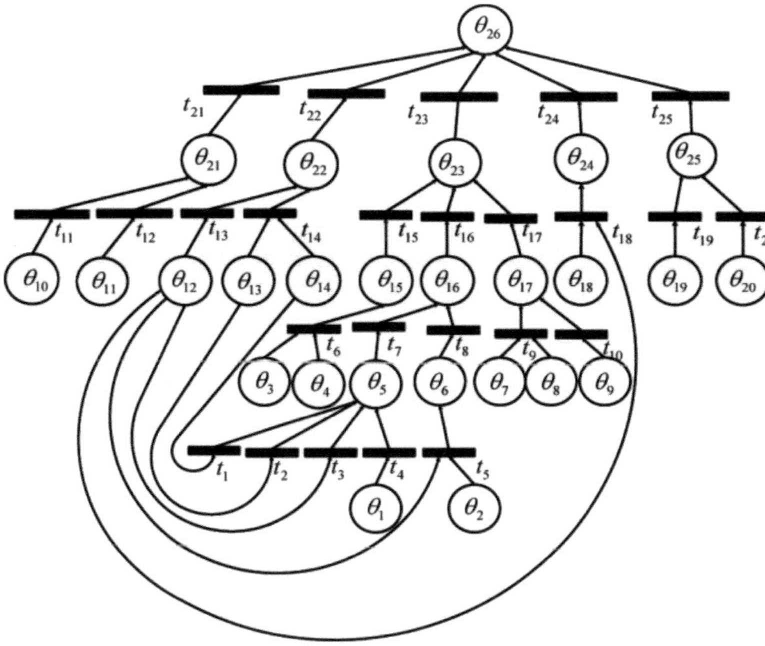

Figure 7. The FRPN model of the solar array for reliability apportionment.

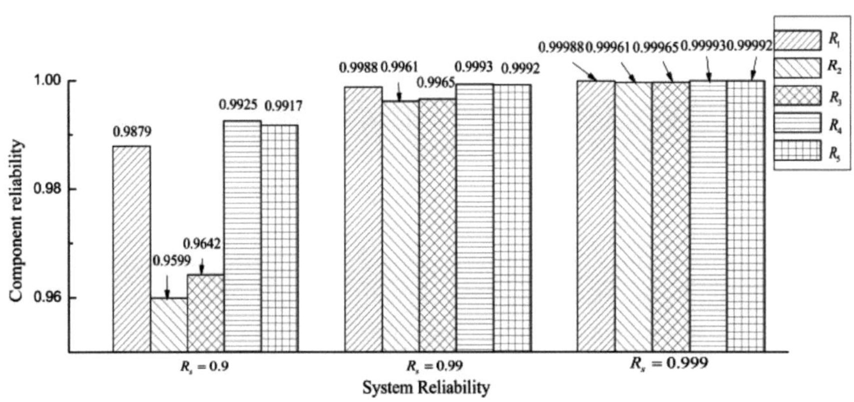

Figure 8. The reliability apportionment of the five key components of solar array. The reliability system is equal to 0.9, 0.99 and 0.999 respectively.

Marker	Event	Truth degree	Marker	Event	Truth degree
P_1	Grease used in hinges between panels	0.4	P_{14}	Particles in space	—
P_2	Brass gasket	0.5	P_{15}	Driving mechanism	—
P_3	Main deriving torsion spring	0.6	P_{16}	Deployable mechanism	—
P_4	Reserved driving torsion spring	0.6	P_{17}	Locking mechanism	—
P_5	Driving pin in the hinge	—	P_{18}	Steel wire	0.7
P_6	Side wall of the hinge	—	P_{19}	Stepping motor	0.2
P_7	Main locking spring	0.8	P_{20}	Transmission system	0.6
P_8	Reserved locking spring	0.5	P_{21}	Hold-down and release mechanism	—
P_9	Locking pin of the hinge	0.5	P_{22}	Solar panels	—
P_{10}	Electronic arcing of the hold-down and release mechanism	0.7	P_{23}	Hinges	—
P_{11}	Cutter of the of the hold-down and release mechanism	0.7	P_{24}	Synchronization mechanism	—
P_{12}	Harsh thermal environment in space	0.9	P_{25}	Orientation mechanism	—
P_{13}	Vacuum and micro-gravity environment in space	—	P_{26}	Mechanical system of the solar array	—

Table 2. Markers and events of FRPN model for reliability apportionment.

3.2.2. Case study

We take the spacecraft solar array as an example to conduct the reliability apportionment by using the FRPN model (**Figure 3**). According to the operational principle of array mechanical systems of a solar array, we build an FRPN model for reliability apportionment of spacecraft solar array. The graphical representation of this model is shown in **Figure 7**. **Table 2** shows the markers and events of the FRPN model [32].

From **Figure 7**, the FRPN model of solar array includes 13 bottom places- P_1, P_2, P_3, P_4, P_7, P_8, P_9, P_{10}, P_{11}, P_{12}, P_{18}, P_{19} and P_{20}. And P_{21}, P_{22}, P_{23}, P_{24}, and P_{25} represent the subsystems (**Table 2**). The final reliability apportionment results are illustrated in **Figure 8** under the system reliability of 0.9, 0.99 and 0.999. In this figure, R_S represents the reliability of the system and $R_i(i = 21, 22, 23, 24, 25)$ expresses the reliability of the five key subsystems. The reliability apportionments are shown in **Figure 8**. By using the FRPN based model, the system reliability can be allocated considering the environmental factors and the intrinsic connection in the mechanical system itself [33].

3.3. Reliability analysis by FRPN

3.3.1. Method

Reliability analysis happens in the stage that the mechanical system has been built physically. By using the FRPN model, we can analyze the reliability of the system with the following steps:

1. Decompose the mechanical system.
2. Build the FRPN model of the mechanical system.
3. Get the truth degrees of the bottom places according to the characteristics of the faults in the system, operation data and engineering experience
4. Calculate the truth degree of top place.
5. Use the cosine matching function (CMF) to analyze reliability of the system.

3.3.2. Case study

We also take the spacecraft solar array as a case for reliability analysis. **Figure 9** shows the FRPN model of the spacecraft solar array for reliability analysis and **Table 3** represents markers and events [37].

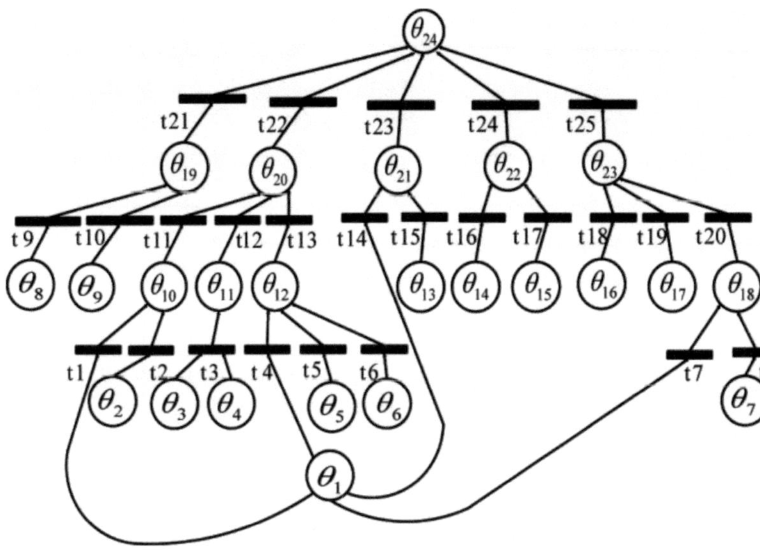

Figure 9. The FRPN model of solar array for reliability analysis.

Marker	Event	Marker	Event
P_{24}	Failure of the solar array system	P_1	Harsh thermal environment in space
P_{19}	Fault of the unlock-mechanism	P_2	Fault of the grease used in hinges between panels
P_{20}	Faults during deployment process	P_3	Insufficient torque of the main torsion spring
P_{21}	Faults during locking process	P_4	Insufficient torque of the reserved torsion spring
P_{22}	Fault of orientation to the sun	P_5	Insufficient preload of the cable
P_{23}	Other faults of mechanical system	P_6	Poor thermal characteristic of the cable
P_{10}	Deadlocking in hinges	P_{13}	Inappropriate driving torque of the locking torsion spring
P_{11}	Insufficient preload of the torsion spring	P_{14}	Fault of the motor

Marker	Event	Marker	Event
P_{12}	Fault of CCL	P_{15}	Fault of the transmission unit
P_{18}	Vibration of panels induced by thermal deformation	P_{16}	Impact caused by particles in space
P_8	Electronic arcing is out of service	P_{17}	Vibration caused by clearances of hinges
P_9	Fault of the cutters	P_7	Bad thermal characteristic of honeycomb materials

Table 3. Markers and events of FRPN for reliability analysis.

Define θ_i as the truth degree of the bottom place p_i, $\theta_i \in [0, 1]$. A higher value indicates that the possibility of the event is higher, which means the fault occurs much easier. **Table 4** demonstrates the ranks, occurrence, and truth degrees of the bottom places. According to the characteristics of

Rank	I	II	III	IV	V	VI	VII
Occurrence	Very low	Low	Fairly low	Moderate	Fairly high	High	Very high
Truth degree	0.1	0.3	0.4	0.5	0.6	0.8	1.0

Table 4. Solar array classification ranks of the fault model.

Marker of bottom places	P_1	P_2	P_3	P_4	P_5	P_6	P_7
Rank	VII	III	V	V	VI	V	VI
Truth degree	1.0	0.4	0.6	0.4	0.8	0.6	0.8
Marker of bottom places	P_8	P_9	P_{13}	P_{14}	P_{15}	P_{16}	P_{17}
Rank	V	V	V	II	IV	VI	VI
Truth degree	0.4	0.6	0.8	0.3	0.5	0.8	0.8

Table 5. Fault rank of the bottom places and their truth degree.

Bottom place	Improvement measures
P_1	The thermal environment in space is the crucial factor of the failure. Some approaches to improve the reliability of the system. (1) Investigate the temperature in space precisely where the solar array works and sum the rules; (2) use new material that is fit for the change of the temperature in space; (3) research the temperature impact on the structure, and optimize the structure of the crucial part of the system
P_{13}	(1) Test the torsion spring on the ground, then find the torque-angle curve to know the characteristics of the torsion spring more deeply; (2) test the performance of the whole system, using torsion springs with different characters, like stiffness
P_{16}	That happens occasionally. There is no effective measure to avoid particles in space, maybe only two ways: (1) make the structure stronger; (2) make the system more agile to detect the vibration caused by the impact of particles, and make adjustment with expedition

Table 6. Improvement measures.

the faults in the system, operation data and engineering experience [9]. **Table 5** represents the fault rank of the bottom places and their truth degrees.

We can get the results of reliability analysis by using the method in Section 3.3.1. According to the results, we can evaluate the importance of bottom places in the FRPN model. Some details can be checked in [37]. To improve the system reliability, we should propose some approaches to enhance the weak links. **Table 6** shows some improvement measures for the mechanical system of a spacecraft solar array.

4. Conclusion

With the ever-increased high requirement of reliability and safety for critical equipment, accurately performing the reliability evaluation of the mechanical systems, such as solar arrays, gains much attention in recent years. The proposed method for reliability evaluation by FRPN can be used to solve the problem on how to describe the relationship among the different components and how to overcome data deficiency. The FRPN based models may open up a new way for evaluating complex mechanical systems with multi-state operation in variable working environment.

Acknowledgements

This work was supported by the National Science Foundation of China under Contract No. 50875149, High Technology Project under Contract No. 2009AA04Z401, and Research Project of The State Key Laboratory of Tribology under Contract No. SKLT11B03.

Conflict of interest

There has no conflict of interests.

Author details

Jianing Wu[1,2]* and Shaoze Yan[1]

*Address all correspondence to: jianing.wu@me.gatech.edu

1 Division of Intelligent and Biomechanical Systems, State Key Laboratory of Tribology, Department of Mechanical Engineering, Tsinghua University, Beijing, China

2 School of Mechanical Engineering, Georgia Institute of Technology, Atlanta, GA, USA

References

[1] Henley EJ, Kumamoto H. Reliability Engineering and Risk Assessment. Vol. 568. Englewood Cliffs, NJ: Prentice-Hall; 1981

[2] Brandhorst HW Jr, Rodiek JA. Space solar array reliability: A study and recommendations. Acta Astronautica. 2008;**63**(11–12):1233-1238. DOI: 10.1016/j.actaastro.2008.05.010

[3] Harland DM, Lorenz RD. Space Systems Failures. 1st ed. Chichester: Springer-Praxis; 2005. DOI: 10.1007/978-0-387-27961-9

[4] Huang W, Askin RG. Reliability analysis of electronic devices with multiple competing failure modes involving performance aging degradation. Quality and Reliability Engineering International. 2003;**19**(3):241-254. DOI: 10.1002/qre.524

[5] Arifujjaman M, Iqbal MT, Quaicoe JE. Reliability analysis of grid connected small wind turbine power electronics. Applied Energy. 2009;**86**(9):1617-1623. DOI: 10.1016/j.apenergy.2009.01.009

[6] Foucher B, Boullie J, Meslet B, Das D. A review of reliability prediction methods for electronic devices. Microelectronics Reliability. 2002;**42**(8):1155-1162. DOI: 10.1016/S0026-2714(02)00087-2

[7] Rausand M, Hφyland A. System Reliability Theory: Models, Statistical Methods, and Applications. 2nd ed. Hoboken: John Wiley & Sons, Inc.; 2004. DOI: 10.1198/tech.2004.s242

[8] Bertsche B. Reliability in Automotive and Mechanical Engineering: Determination of Component and System Reliability. Hoboken, New Jersey: Springer Science & Business Media; 2008

[9] Tang J. Mechanical system reliability analysis using a combination of graph theory and Boolean function. Reliability Engineering & System Safety. 2001;**72**(1):21-30

[10] Zeng SK, Zhao TD, Zhang JG, et al. Design and Analysis of System Reliability. Beijing: Beijing University of Aeronautics and Astronautics Press; 2001. (in Chinese)

[11] Son YK. Reliability prediction of engineering systems with competing failure modes due to component degradation. Journal of Mechanical Science and Technology. 2010;**25**: 1717-1725

[12] Hu Y, Zhu MR. Handbook of Reliability Design. Beijing: China Zhijian Publishing House; 2007. (in Chinese)

[13] Huang HZ, Qu J, Zuo MJ. Genetic-algorithm-based optimal apportionment of reliability and redundancy under multiple objectives. IIE Transactions. 2009;**41**(4):287-298. DOI: 10.1080/07408170802322994

[14] James KB, Donald HG. Reliability growth apportionment. IEEE Transactions on Reliability. 1977;**26**(4):242-244. DOI: 10.1109/TR.1977.5220138

[15] Park KS. Fuzzy apportionment of system reliability. IEEE Transactions on Reliability. 1987; **36**(1):129-132. DOI: 10.1109/TR.1987.5222317

[16] Dhingra AK. Optimal apportionment of reliability and redundancy in series systems under multiple objectives. IEEE Transactions on Reliability. 1992;**41**(4):576-582. DOI: 10.1109/24.249589

[17] Mentes A, Helvacioglu IH. An application of fuzzy fault tree analysis for spread mooring systems. Ocean Engineering. 2011;**38**:285-294. DOI: 10.1016/j.oceaneng.2010.11.003

[18] Lee WS, Grosh DL, Tillman FA, Lie CH. Fault tree analysis, methods, and applications: A review. IEEE Transactions on Reliability. 1985;**34**(3):194-203. DOI: 10.1109/TR.1985.5222114

[19] de Queiroz Souza R, Álvares AJ. FMEA and FTA analysis for application of the reliability centered maintenance methodology: Case study on hydraulic turbines. In: ABCM Symposium Series in Mechatronics; Vol. 3; 2008. pp. 803-812

[20] Gao M, Zhou M, Huang X, Wu Z. Fuzzy reasoning Petri nets. IEEE Transactions on Systems, Man, and Cybernetics - Part A: Systems and Humans. 2003;**33**(3):314-324. DOI: 10.1109/TSMCA.2002.804362

[21] Murata T. Petri nets: Properties, analysis and applications. Proceedings of the IEEE. 1989; **77**(4):541-580. DOI: 10.1109/5.24143

[22] Yuan CY. Petri Nets. 1st ed. Nanjing: Press of Southeastern University; 1989. (in Chinese)

[23] Pedrycz W, Gomide F. A generalized fuzzy Petri net model. IEEE Transactions on Fuzzy Systems. 1994;**2**(4):295-301. DOI: 10.1109/91.324809

[24] Uzam M. The use of the Petri net reduction approach for an optimal deadlock prevention policy for flexible manufacturing systems. International Journal of Advanced Manufacturing Technology. 2004;**23**(3–4):204-219. DOI: 10.1007/s00170-002-1526-5

[25] Zhao JH, Liu ZH, Dao MT. Reliability optimization using multiobjective ant colony system approaches. Reliability Engineering and System Safety. 2007;**92**(1):109-120. DOI: 10.1016/j.ress.2005.12.001

[26] Leveson NG, Stolzy JL. Safety analysis using Petri nets. IEEE Transactions on Software Engineering. 1987;**13**(3):386-397. DOI: 10.1109/TSE.1987.233170

[27] Yang BS, Jeong SK, Oh YM, Tan ACC. Case-based reasoning system with Petri nets for induction motor fault diagnosis. Expert Systems with Applications. 2004;**27**(2):301-311. DOI: 10.1016/j.eswa.2004.02.004

[28] Constantinescu C. Trends and challenges in VLSI circuit reliability. IEEE Micro. 2003;**23**(4): 14-19. DOI: 10.1109/MM.2003.1225959

[29] Wallrapp O, Wiedemann S. Simulation of deployment of a flexible solar array. Multibody System Dynamics. 2002;**7**(1):101-125. DOI: 10.1023/A:1015295720991

[30] Rauschenbach HS. Solar Cell Array Design Handbook: The Principles and Technology of Photovoltaic Energy Conversion. 1st ed. Beijing: China Astronautic Publishing House; 1994. (in Chinese)

[31] Fragnito M, Pastena M. Design of smart microsatellite deployable solar wings. Acta Astronautica. 2000;**46**(2–6):335-344. DOI: 10.1016/S0094-5765(99)00228-3

[32] Wu J, Yan S. An approach to system reliability prediction for mechanical equipment using fuzzy reasoning Petri net. Proceedings of the Institution of Mechanical Engineers, Part O: Journal of Risk and Reliability. 2014;**228**(1):39-51. DOI: 10.1177/1748006X13495130

[33] Wu J, Yan S, Xie L, Gao P. Reliability apportionment approach for spacecraft solar array using fuzzy reasoning Petri net and fuzzy comprehensive evaluation. Acta Astronautica. 2012;**76**:136-144. DOI: 10.1016/j.actaastro.2012.02.023

[34] Castet JF, Saleh JH. Satellite and satellite subsystems reliability: Statistical data analysis and modeling. Reliability Engineering & System Safety. 2009;**94**(11):1718-1728. DOI: 10.1016/j.ress.2009.05.004

[35] Rome Laboratory. Reliability Prediction of Electronic Equipment. 1991. Available from: www.barringer1.com/mil_files/MIL-HDBK-217RevF.pdf [Accessed: October 24, 2012]

[36] USAF Rome Air Development Center. Non-Electronics Parts Reliability Data (NPRD). 1995. Available from: www.theriac.org/riacapps/search/?category=all%20products&keyword=nprd [Accessed: October 24, 2012]

[37] Wu J, Yan S, Xie L. Reliability analysis method of a solar array by using fault tree analysis and fuzzy reasoning Petri net. Acta Astronautica. 2011;**69**(11–12):960-968. DOI: 10.1016/j.actaastro.2011.07.012

Chapter 5

Performance Analysis of Shared-Memory Bus-Based Multiprocessors Using Timed Petri Nets

Wlodek M. Zuberek

Additional information is available at the end of the chapter

http://dx.doi.org/10.5772/intechopen.75589

Abstract

In shared-memory bus-based multiprocessors, the number of processors is often limited by the (shared) bus; when the utilization of the bus approaches 100%, processors spend an increasing amount of time waiting to get access to the bus (and shared memory) and this degrades their performance. The limitations imposed by the bus depend upon many parameters, and different parameters affect the performance in different ways. This chapter uses timed Petri nets to model shared-memory bus-based multiprocessors at the instruction execution level and shows how the performance of processors and the system are affected by different modeling parameters. Discrete-event simulation of the developed net models is used to get performance results.

Keywords: shared-memory multiprocessors, bus-based multiprocessors, timed Petri nets, performance analysis, discrete-event simulation

1. Introduction

Due to continuous progress in manufacturing technologies, the performance of microprocessors has been steadily improving over several decades, doubling every 18 months (the so-called Moore's law [1]). The capacity of memory chips has also been doubling every 18 months, but the performance has been improving less than 10% per year [2]. The performance gap [3] between the processor and its memory have been doubling approximately every 6 years, and an increasing part of the processor's time is being spent on waiting for the completion of memory operations. Although multilevel cache memories are used to reduce the average latencies of memory accesses, matching the performances of the processor and the memory is an increasingly difficult task. In effect, it is often the case that more than 50% of processor cycles are spent waiting for the completion of memory accesses [4]. A model of a pipelined processor at the instruction execution level is used in this chapter to study the mismatch of processor and memory performances.

This model of a processor is then used for performance analysis of shared-memory bus-based multiprocessors. The main objective of this analysis is to study the degradation of the processor's performance when the utilization of the (shared) bus approaches 100%. This performance degradation limits the number of processors in bus-based systems.

Modeling and analysis of shared-memory bus-based systems requires a flexible formalism that can easily handle concurrent activities as well as synchronization of different events and processes that occur in such systems. Petri nets [5, 6] are such formal models.

As formal models, Petri nets are bipartite directed graphs in which the two types of vertices represent, in a very general sense, conditions and events. An event can occur only when all conditions associated with it (represented by arcs directed to the event) are satisfied. An occurrence of an event usually satisfies some other conditions, indicated by arcs directed from the event. Hence, an occurrence of one event causes some other event (or events) to occur, and so on.

In order to study the performance aspects of systems modeled by Petri nets, the durations of modeled activities must also be taken into account. This can be done in different ways, resulting in different types of temporal nets [7]. In timed Petri nets [8], occurrence times are associated with events, and the events occur in real time (as opposed to instantaneous occurrences in other models). For timed nets with constant or exponentially distributed occurrence times, the state graph of a net is a Markov chain (or an embedded Markov chain). If the state space of a timed net is finite and reasonably small, the stationary probabilities of states can be determined by standard methods [9]. Then these stationary probabilities are used for the derivation of many performance characteristics of the model [10]. In other cases, discrete event simulation [11] is used to find the performance characteristics of a timed net.

In this chapter, timed Petri nets are used to model shared-memory bus-based multiprocessor systems. Section 2 recalls basic concepts of Petri nets and timed Petri nets. Section 3 discusses a model of a pipelined processor and its performance as a function of modeling parameters. Shared-memory bus-based systems are described and analyzed in Section 4. Section 5 concludes the chapter.

2. Timed Petri nets

In Petri nets, concurrent activities are represented by tokens that can move within a (static) graph-like structure of the net. More formally, a marked place/transition Petri net $\mathcal{M}$ is defined as a pair $\mathcal{M} = (\mathcal{N}, m_0)$, where the structure N is a bipartite directed graph, $\mathcal{N} = (P, T, A)$, with two types of vertices, a set of places P and a set of transitions T, and a set of directed arcs A connecting places with transitions and transitions with places, $A \subseteq P \times T \cup T \times P$. The initial marking function m_0 assigns nonnegative numbers of tokens to places of the net, $m_0 : P \rightarrow \{0, 1, \ldots\}$. Marked nets can be equivalently defined as $\mathcal{M} = (P, T, A, m_0)$.

A place is shared if it is connected to more than one transition. A shared place p is free-choice if the sets of places connected by directed arcs to all transitions sharing p are identical. A shared place p is (dynamically) conflict-free if for each marking reachable from the initial marking, at most one transition sharing p is enabled. If a shared place p is not free-choice and not conflict-free, the transitions sharing p are conflicting.

In timed nets [8], occurrence times are associated with transitions, and transition occurrences are real-time events, i.e., tokens are removed from input places at the beginning of the occurrence period and are deposited to the output places at the end of this period. All occurrences of enabled transitions are initiated in the same instants of time in which the transitions become enabled (although some enabled transitions may not initiate their occurrences). If, during the occurrence period of a transition, the transition becomes enabled again, a new, independent occurrence can be initiated, which will overlap with the other occurrence(s). There is no limit on the number of simultaneous occurrences of the same transition (sometimes this is called infinite occurrence semantics). Similarly, if a transition is enabled "several times" (i.e., it remains enabled after initiating an occurrence), it may start several independent occurrences in the same time instant.

More formally, a timed Petri net is a triple, $\mathcal{T} = (\mathcal{M}, c, f)$, where $\mathcal{M}$ is a marked net, c is a choice function which assigns probabilities to transitions in free-choice classes, or relative frequencies of occurrences to conflicting transitions, $c \rightarrow [0, 1]$, and f is a timing function, which assigns an (average) occurrence time to each transition of the net, $f : T \rightarrow \mathbf{R}^+$, where $\mathbf{R}^+$ is the set of nonnegative real numbers.

The occurrence times of transitions can be either deterministic or stochastic (i.e., described by some probability distribution function). In the first case, the corresponding timed nets are referred to as D-timed nets [12]; in the second, for the (negative) exponential distribution of occurrence times, the nets are called M-timed nets (Markovian nets) [13]. In both cases, the concepts of state and state transitions have been formally defined and used in the derivation of different performance characteristics of the model. In simulation applications, other distributions can also be used, for example, the uniform distribution (U-timed nets) is sometimes a convenient option. In timed Petri nets, different distributions can be associated with different transitions in the same model providing flexibility that is used in simulation examples that follow.

In timed nets, the occurrence times of some transitions may be equal to zero, which means that the occurrences are instantaneous; all such transitions are called immediate (while the others are called timed). Since immediate transitions have no tangible effects on the (timed) behavior of the model, it is convenient to "split" the set of transitions into two parts, the set of immediate and the set of timed transitions, and to first perform all occurrences of the (enabled) immediate transitions, and then (still in the same time instant), when no more immediate transitions are enabled, to start the occurrences of (enabled) timed transitions. It should be noted that such a convention effectively introduces the priority of immediate transitions over the timed ones, so the conflicts of immediate and timed transitions are not allowed in timed nets. Detailed characterization of the behavior of timed nets with immediate and timed transitions is given in [8].

3. Pipelined processors

A timed Petri net model of a pipelined processor [14] at the level of instruction execution is shown in **Figure 1** (as usual, timed transitions are represented by solid bars, and immediate transitions by thin bars). It is assumed that the first level cache does not delay the processor, while cache misses (at the first level cache) introduce a delay of t_c processor cycles. For simplicity, only two levels of cache memory are represented in the model; it appears that such a simplification does not affect the results in a significant way [15].

Place *Pnxt* is marked when the processor is ready to execute the next instruction. *Pnxt* is a free-choice place with three possible outcomes that model issuing an instruction without any further delay (*Ts0* with the choice probability p_{s0}), a single-cycle pipeline stall (modeled by *Td1* with the choice probability p_{s1} associated with *Ts1*), and a two-cycle pipeline stall (modeled by *Td2* and then *Td1* with the choice probability p_{s2} assigned to *Ts2*). Other pipeline stalls could be represented in a similar way, if needed.

Marked place *Cont* indicates that an instruction is ready to be issued to the execution pipeline. It is assumed that once the instruction enters the pipeline, it will progress through all the stages and, eventually, leave the pipeline. Since the details of pipeline implementation are not important for performance analysis of the processor, they are not represented here. Only the first stage of the execution pipeline is shown as timed transition *Trun*.

Done is another free-choice place which determines if the executing instruction results in a cache miss or not. Transition *Tnxt* occurs (with the corresponding probability) if cache miss does not occur and the processor can continue fetching and issuing instructions. Cache miss is represented by *Tsel*. The choice probability associated with *Tsel* determines the instruction

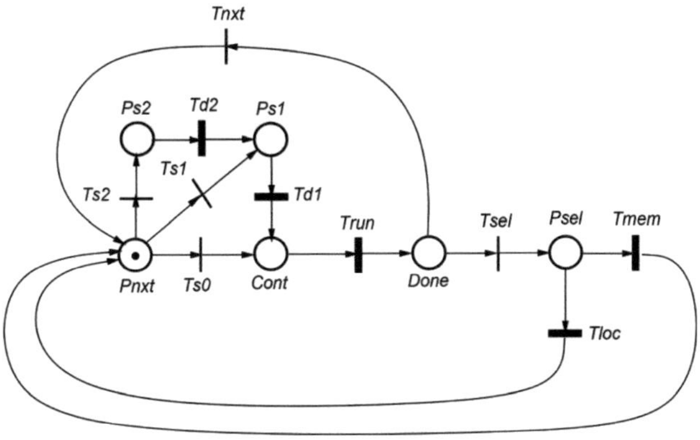

Figure 1. Instruction-level Petri net model of a pipelined processor.

runlength, n_ℓ, i.e., the average number of instructions between two consecutive cache misses; if this choice probability is equal to 0.1, the runlength is equal to 10; if it is equal to 0.2, the runlength is 5; and so on.

Psel is another free-choice place; it models the hits and misses of the second-level cache. The probability associated with transition *Tloc* represents the hit ratio of the second-level cache (the occurrence time of *Tloc* is the average access time to the second-level cache, t_c) while the miss ratio is associated with transition *Tmem* which represents accesses to the main memory (with the occurrence time t_m).

Typical values of modeling parameters used in this chapter are shown in **Table 1**.

All temporal data in **Table 1** (i.e., cache and memory access times) are in processor cycles.

Processor utilization as a function of h_1, the hit rate of the first-level cache, is shown in **Figure 2** for two values of the second-level cache access time, $t_c = 5$, and $t_c = 10$. It should not be

Symbol	Parameter	Value
h_1	First-level cache hit rate	0.9
h_2	Second-level cache hit rate	0.8
t_p	First-level cache access time	1
t_c	Second-level cache access time	5
t_m	Main memory access time	25
p_{s1}	Prob. of one-cycle pipeline stall	0.1
p_{s2}	Prob. of two-cycle pipeline stall	0.05

Table 1. Modeling parameters and their typical values.

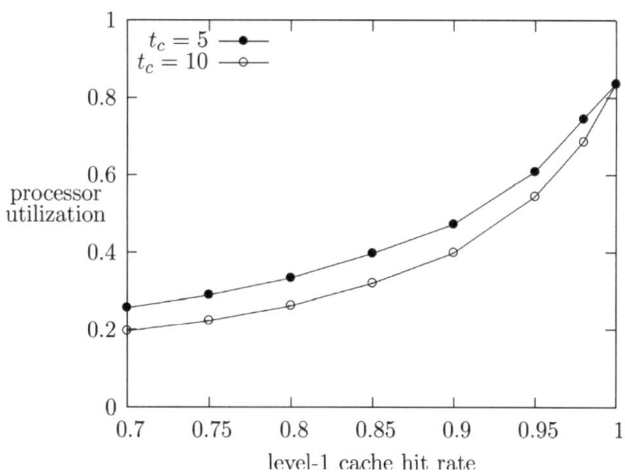

Figure 2. Processor utilization as a function of first-level cache hit rate for $h_2 = 0.8, p_s = 0.2$.

surprising that processor utilization is quite sensitive to the values of h_1, but is much less sensitive to the values of t_c.

Processor utilization as a function of h_2, the hit rate of the second-level cache, is shown in **Figure 3** for two values of the main memory access time, $t_m = 25$ and $t_m = 50$. Processor utilization is rather insensitive to values of h_2, and does not change much with t_m.

Processor utilization as a function of the probability of pipeline stalls $p_s = p_{s1} + 2p_{s2}$ is shown in **Figure 4** for three combinations of values of t_c and t_m.

Again, processor utilization is rather insensitive to the probability of pipeline stalls as well as the values of t_c and t_m.

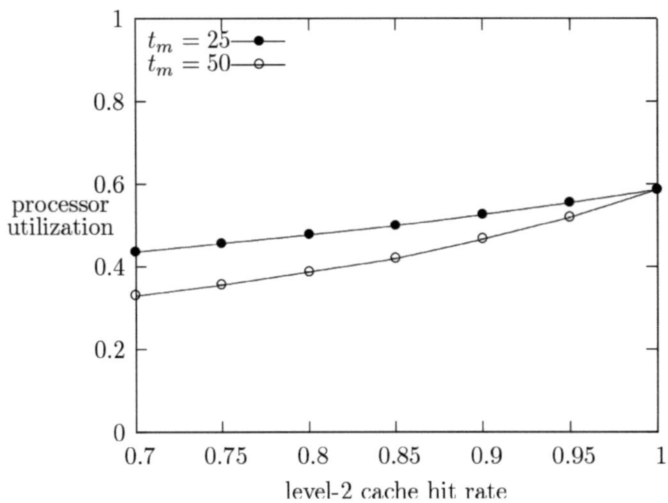

Figure 3. Processor utilization as a function of second-level cache hit rate for $h_1 = 0.9, p_s = 0.2$.

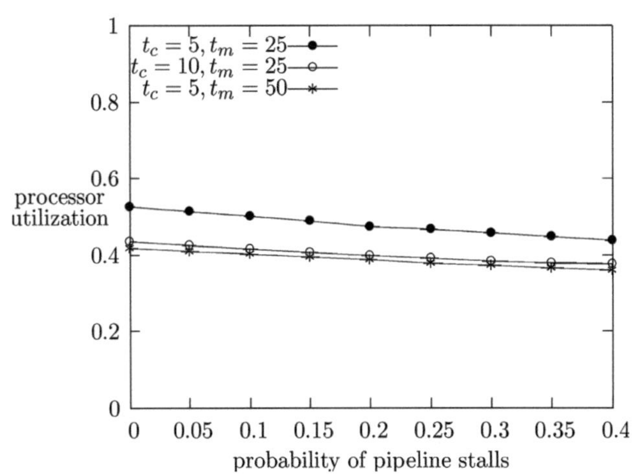

Figure 4. Processor utilization as a function of probability of pipeline stalls for $h_1 = 0.9, h_2 = 0.8$.

For pipelined processors shown in **Figure 1**, processor utilization can be estimated using the following formula:

$$u_p = \frac{1}{1 + p_{s1} + 2{*}p_{s2} + (1 - h_1){*}(t_c + (1 - h_2){*}t_m)}. \tag{1}$$

For the values of modeling parameters shown in **Table 1**, processor utilization is:

$$u_p = \frac{1}{1 + 0.1 + 0.1 + 0.1{*}(5 + 0.2{*}25)} \approx 0.45. \tag{2}$$

The estimated values agree quite well with the values shown in **Figures 2–4**.

4. Shared-memory bus–based systems

An outline of a shared-memory bus-based multiprocessor is shown in **Figure 5**. The system is composed of n identical processors, which access the shared memory using a system bus. To reduce the average access time to the shared memory, the processors use (multilevel) cache memories. It is assumed that memory consistency is provided by a cache coherence mechanism [16], which usually increases the miss ratio of accessing caches (and is otherwise not represented in the model).

A timed Petri net model of a shared-memory bus-based multiprocessor is shown in **Figure 6**. It contains models of n processors (only two are shown in **Figure 6**), which are copies of the model shown in **Figure 1** except for the main memory (transition *Tmem*) which becomes shared memory in **Figure 6**. The remaining part of **Figure 6** is modeling the bus that coordinates accesses of processors to the shared memory.

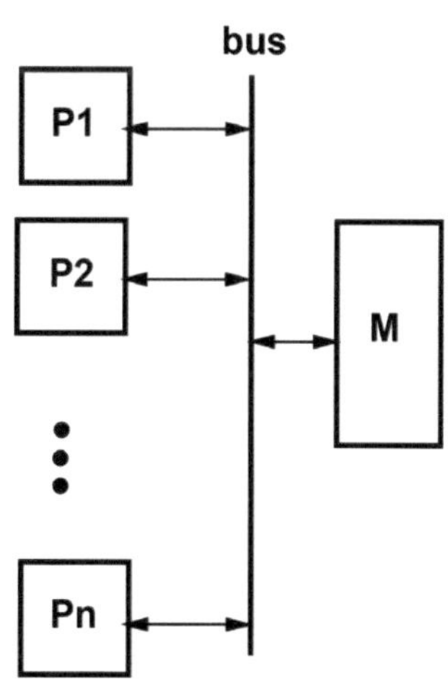

Figure 5. A shared-memory bus-based multiprocessor.

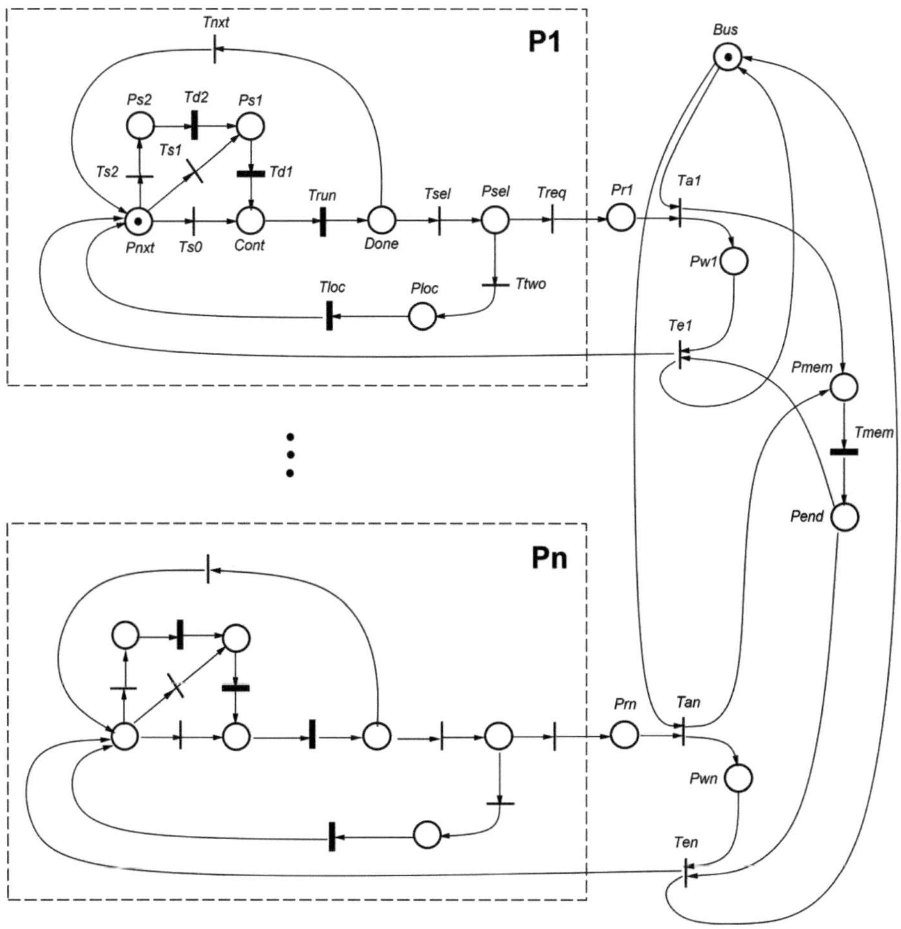

Figure 6. A timed Petri net model of bus-based shared-memory multiprocessor.

When a processor i, $i = 1, \ldots, n$, requests an access to shared memory, place *Pri* becomes marked. If the bus is available (i.e., if place *Bus* is marked), the occurrence of transition *Tai* indicates that processor *i* begins its access to shared memory. Transitions *Ta*1, ..., *Tan* constitute a conflict class with a fair resolution of conflicts (i.e., all conflicting processors have the same probability of being selected for accessing memory). In real systems, accessing the shared bus is often based on priorities assigned to processors; such priorities could easily be represented using inhibitor arcs in Petri nets.

Place *Pmem* collects memory access requests from all processors (occurrences of transitions *Tai*). The end of memory access (i.e., the end of the occurrence of *Tmem*) is indicated by an occurrence of transition *Tei* of the processor which initiated memory access. The occurrence of *Tei* also returns a token to *Bus*, allowing another access to shared memory to be executed.

Figure 7 shows the utilization of processors and the bus as functions of the number of processors in a shared-memory system for the values of modeling parameters shown in **Table 1**.

http://dx.doi.org/10.5772/intechopen.75589

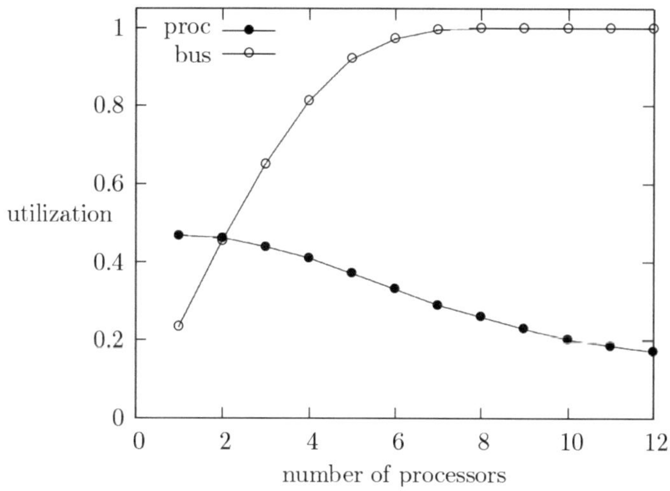

Figure 7. Processor and bus utilization as functions of the number of processors for $h_1 = 0.9$, $h_2 = 0.8$, $p_s = 0.2$.

In **Figure 7**, the bus utilization approaches 100% for about five processors. Moreover, the degradation of processors' performance due to increasing waiting times for accessing the bus (and shared memory) is well illustrated in **Figure 7**.

The average waiting time (in processor cycles) of accessing shared memory (i.e., the times from requesting memory access in place *Pri* to granting this access by an occurrence of *Tai*) is shown in **Figure 8** as a function of the number of processors in the system.

Figure 8 shows that the waiting times increase almost linearly with the number of processors when this number is greater than 5, i.e., when the bus (and shared memory) is utilized in almost 100%.

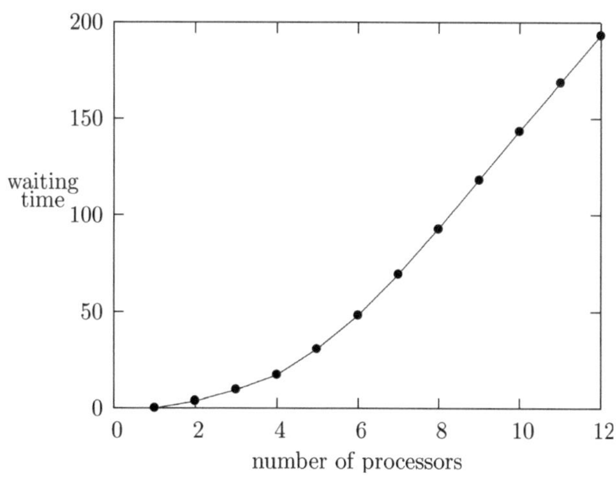

Figure 8. The average waiting time for accessing shared memory as a function of the number of processors for $h_1 = 0.9$, $h_2 = 0.8$, $p_s = 0.2$.

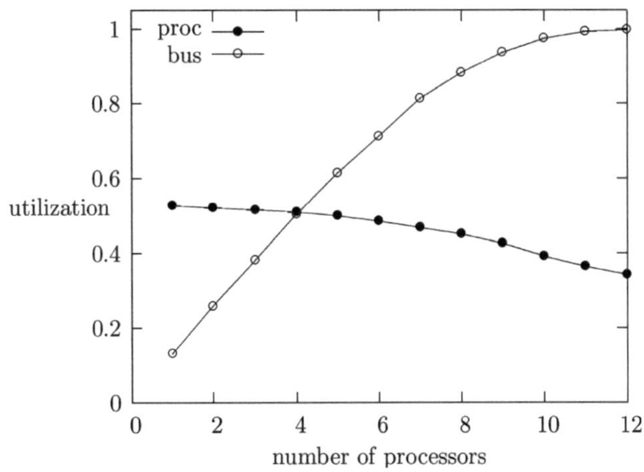

Figure 9. Processor and bus utilization as functions of the number of processors for $h_1 = 0.9$, $h_2 = 0.9$, $p_s = 0.2$.

If the value of the second-level cache hit rate, h_2, increases (and the other parameters do not change), the number of accesses to main memory is reduced, so that the performances of processors and the whole system improve. **Figure 9** shows the utilization of processors and the bus as functions of the number of processors in the system for $h_2 = 0.9$. It also shows that the reduced (in comparison with **Figure 7**) utilization of the bus allows the increase of the number of processors without significant degradation of their performance.

By a similar argument, reduced hit rate at the first-level cache, h_1, increases the number of accesses to the second-level cache as well as to the main memory, and this results in reduced performance of the system. **Figure 10** shows the utilization of processors and the bus as functions of the number of processors in the system for $h_1 = 0.8$. It provides a good illustration of the degradation of performance when compared with **Figure 7**.

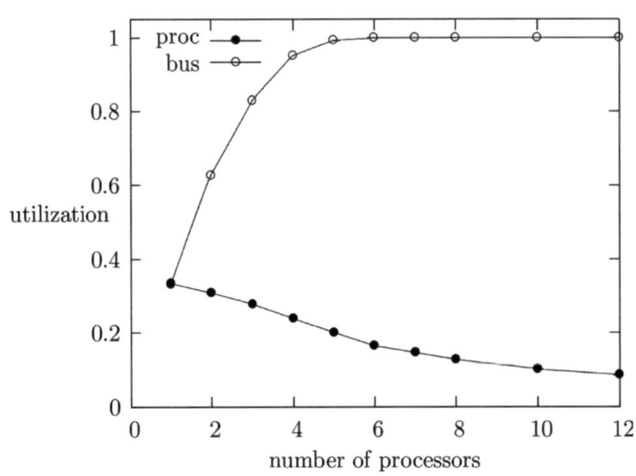

Figure 10. Processor and bus utilization as functions of the number of processors for $h_1 = 0.8$, $h_2 = 0.8$, $p_s = 0.2$.

The number of processors for which the bus is used to almost 100% can be estimated by the following formula:

$$n_p = \frac{1 + p_{s1} + 2{*}p_{s2} + (1 - h_1){*}(t_c + (1 - h_2){*}t_m)}{(1 - h_1){*}(1 - h_2){*}t_m}. \tag{3}$$

For the case shown in **Figure 7**, this number is:

$$\frac{1 + 0.1 + 0.1 + 0.1{*}(5 + 0.2{*}25)}{0.1{*}0.2{*}25} = \frac{1.2 + 1.0}{0.5} = 4.4. \tag{4}$$

For the case shown in **Figure 9**, this value is 8.2.

There are several ways in which the number of processors can be increased in bus-based systems without sacrificing the processors' performance. The simplest approach is to introduce the second bus which allows two concurrent accesses to shared memory, provided the memory is dual port (it allows two concurrent accesses). **Figure 11** outlines a dual bus shared-memory system.

Petri net model of a dual bus system is the same as in **Figure 6**; the only difference is the initial marking of place *Bus*, which now requires two tokens to represent two concurrent accesses to shared memory. It should be observed observed that, for a small number of processors, the utilization of each bus in **Figure 12** is one half of that in **Figure 7**, and also the number of processors that can be used in such a dual bus system without degradation of their performance is twice as large as in a single bus system (**Figure 7**).

The results shown in **Figure 12** are very similar to those shown in **Figure 9**. The second bus in a shared-memory system allows to perform two concurrent accesses to the shared memory. From a single processor's performance point of view, this effect is similar to reducing two

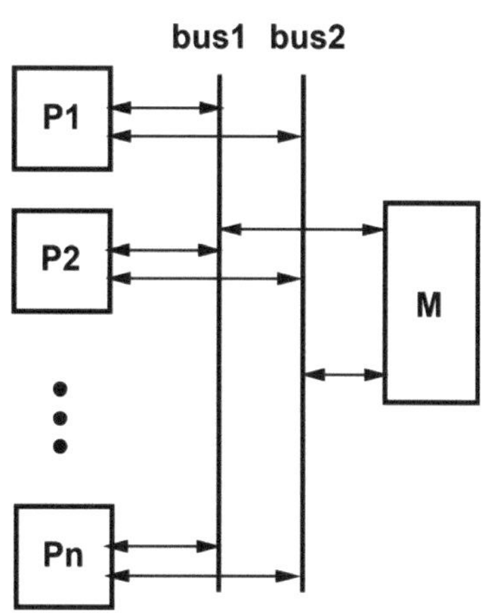

Figure 11. A dual bus shared-memory multiprocessor.

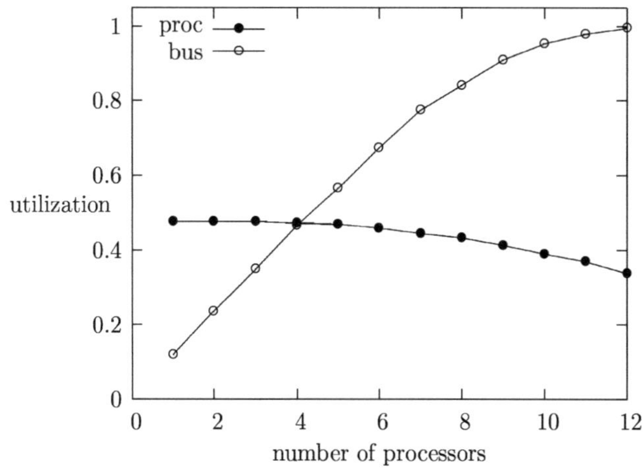

Figure 12. Processor and bus utilization as functions of the number of processors—dual bus system with $h_1 = 0.9$, $h_2 = 0.8$, $p_s = 0.2$.

times the number of accesses to the shared memory, and this is the effect of reducing two times the miss rate for the second-level cache (which is shown in **Figure 9**).

If dual port memory cannot be used, the shared memory can be split into several independent modules which can be accessed concurrently by the processors provided that the bus is also split into sections associated with each module, with processors accessing all such sections, as shown in **Figure 13** for four independent memory modules. The main difference between a multibus system (**Figure 11**) and a system with split bus is in accessing the shared memory; in a multiple bus system, the whole shared memory is accessed by each bus, while in a split bus system (**Figure 13**), each section of the bus accesses only one memory module. In the system

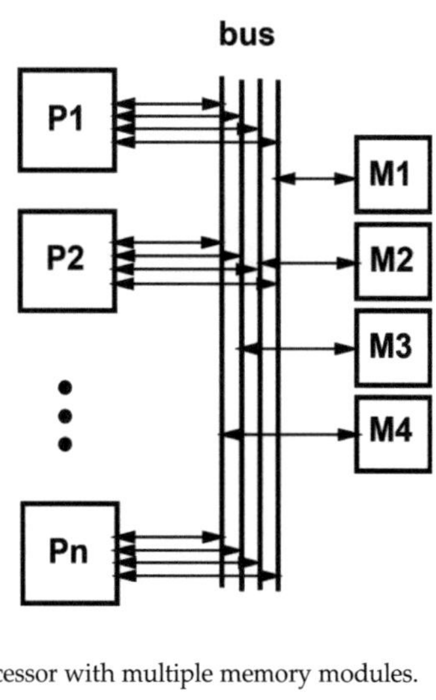

Figure 13. A shared-memory multiprocessor with multiple memory modules.

http://dx.doi.org/10.5772/intechopen.75589

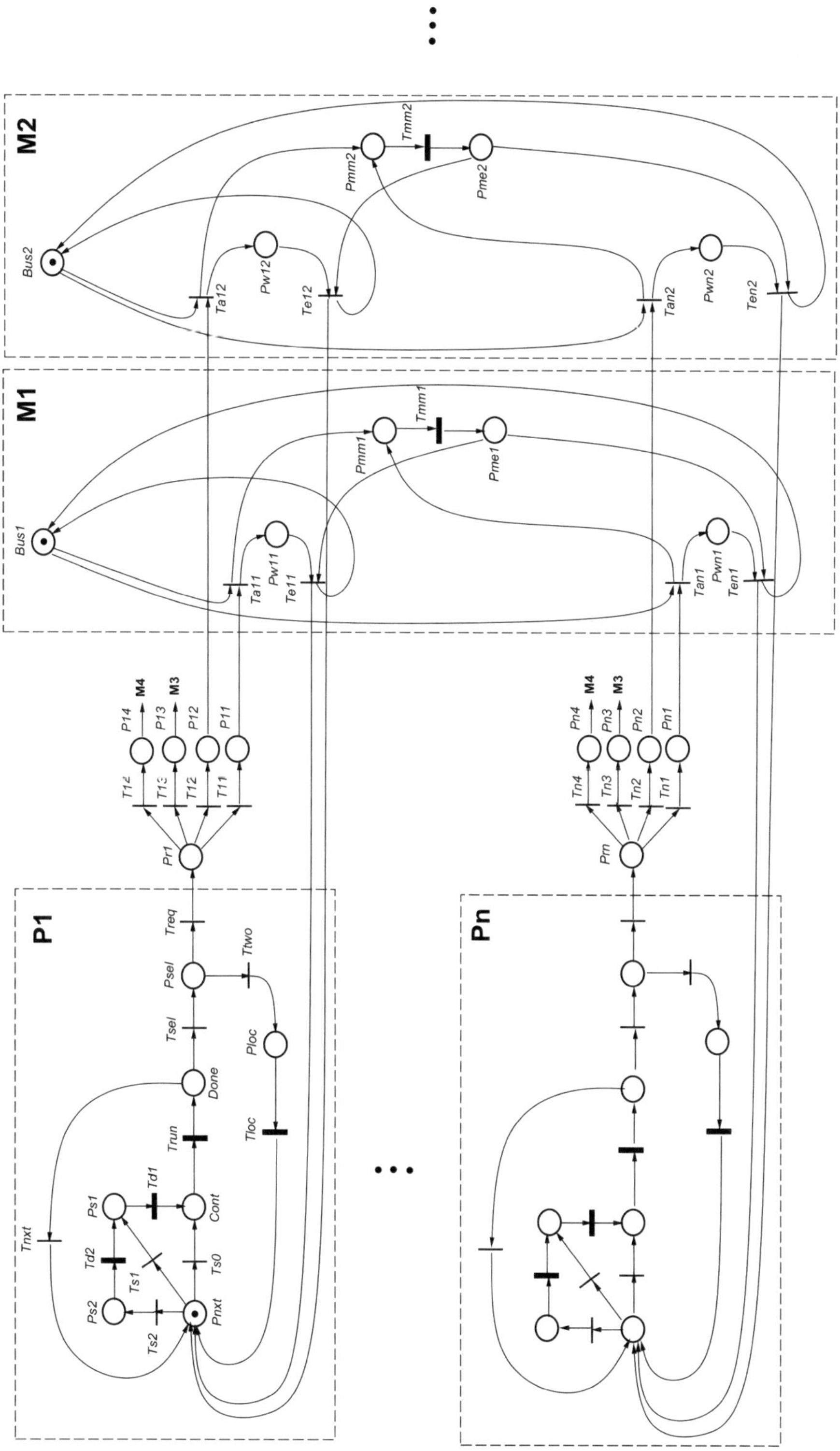

Figure 14. Petri net model of a shared-memory multiprocessor with multiple memory modules.

shown in **Figure 13**, up to four (the number of memory modules) memory accesses can be performed concurrently, but if two (or more) processors request access to the same memory module, the requests are served one after another.

Petri net model of a system outlined in **Figure 13** is shown in **Figure 14** where only two processors and two memory modules are detailed.

In **Figure 14**, there is a free-choice place *Pri* for each processor i, $i = 1, \ldots, n$. This free-choice place selects the requested memory module by transitions *Tij*, $j = 1, \ldots, 4$, and forwards the memory access request to the selected memory module (place *Pij*). If the selected module is available, i.e., if place *Busj* is marked, the access to shared memory is initiated by the occurrence of *Taij*. When this memory access is completed, the occurrence of *Teij* releases the memory modules (by returning a token to *Busj*) and resumes instruction execution in the processor that requested the memory access.

If memory module is not available when it is requested, the memory access is delayed (in *Pij*) until the requested module becomes available.

It is possible that more than one processor becomes waiting for the same memory module. The selection of the processor which will get access first is random with the same probability assigned to all waiting processors. In real systems, there is usually some priority scheme that determines the order in which the waiting processors access the bus. Such a priority scheme could easily be modeled if it is needed (for example, for studying the starvation effect which can be created when the system is overloaded).

In **Figure 14**, the selection of memory modules is random, with the same probabilities for all modules. If this policy is not realistic, a different memory accessing policy can be implemented,

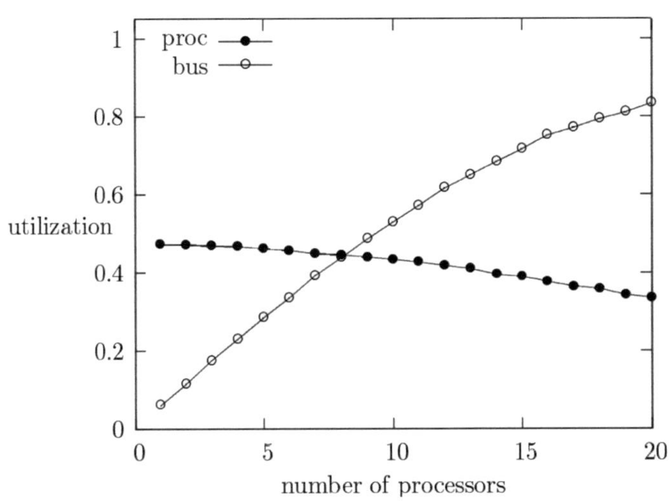

Figure 15. Processor and bus utilization as functions of the number of processors—system with four memory modules and $h_1 = 0.9$, $h_2 = 0.8$, $p_s = 0.2$.

for example, the probabilities of accessing consecutive memory modules by each processor could be used to model sequential processing of large arrays, and so on.

Figure 15 shows the utilization of processors and busses as functions of the number of processors in a system outlined in **Figure 13**.

In **Figure 15**, even for 20 processors, the average utilization of the bus is close to 80%, so the system can accommodate more processors.

5. Concluding remarks

The chapter uses timed Petri nets to model shared-memory bus-based architectures at the level of instruction execution to study the effects of modeling parameters on the performance of the system. The models are rather simple with straightforward representation of modeling parameters.

Performance results presented in this chapter have been obtained by the simulation of developed Petri net models. However, the model shown in **Figure 7** has only 10 states, so its analytical solution (for different values of modeling parameters) can be easily obtained and compared with simulation results to verify their accuracy. **Table 2** shows such a comparison of processor utilization for several combinations of parameters h_1 and h_2. In all cases, the simulation-based results are very close to the analytical ones.

The models of multiprocessor systems are usually composed of many copies of the same submodel of a processor and possibly other elements of the system. Colored Petri nets [17] can significantly simplify such models by eliminating copies of similar subsystems. Analysis of colored Petri nets is, however, much more complex than that of ordinary Petri nets.

Finally, it should be noted that the performance of real-life multiprocessor systems very rarely can be described by a set of parameters that remain stable for any significant period of time. The basic parameters like the hit rates depend upon the executed programs as well as their data, and can change very quickly in a significant way. Consequently, the characteristics presented in this chapter can only be used as some insight into the complex behavior of multiprocessor systems.

h_1	h_2	**Simulated results**	**Analytical results**
0.8	0.8	0.3341	0.3333
0.8	0.9	0.3846	0.3846
0.9	0.8	0.4763	0.4762
0.9	0.9	0.5255	0.5263

Table 2. A comparison of simulation and analytical results.

Author details

Wlodek M. Zuberek

Address all correspondence to: wlodek@mun.ca

Department of Computer Science, Memorial University, Canada

References

[1] Hamilton S. Taking Moore's law into the next century. IEEE Computer. 1999;**32**(1):43-48

[2] Patterson DA, Hennessy JL. Computer Architecture—A Quantitative Approach. 4th ed. San Mateo, CA: Morgan Kaufmann; 2006

[3] Wilkes MV. The memory gap and the future of high-performance memories. ACM Architecture News. 2001;**29**(1):2-7

[4] Mutlu O, Stark J, Wilkerson C, Patt YN. Runahead execution: An effective alternative to large instruction windows. IEEE Micro. 2003;**23**(6):20-25

[5] Murata T. Petri nets: Properties, analysis and applications. Proceedings of IEEE. 1989; **77**(4):541-580

[6] Reisig W. Petri nets—An introduction (EATCS Monographs on Theoretical Computer Science. Vol. 4). New York, NY: Springer-Verlag; 1985

[7] Popova-Zeugmann L. Time and Petri Nets. Berlin, Heidelberg: Springer-Verlag; 2013

[8] Zuberek WM. Timed petri nets—Definitions, properties and applications. Microelectronics and Reliability (Special Issue on Petri Nets and Related Graph Models). 1991;**31**(4):627-644

[9] Allen AA. Probability, Statistics and Queueing Theory with Computer Science Applications. 2nd ed. San Diego, CA: Academic Press; 1991

[10] Jain R. The Art of Computer Systems Performance Analysis. New York, NY: Wiley Interscience; 1991

[11] Pooch UW, Wall JA. Discrete Event Simulation. Boca Raton, FL: CRC Press; 1993

[12] Zuberek WM. D-timed petri nets and modelling of timeouts and protocols. Transactions of the Society for Computer Simulation. 1987;**4**(4):331-357

[13] Zuberek WM. M-timed petri nets, priorities, preemptions, and performance evaluation of systems. In: Advances in Petri Nets 1985 (LNCS 222). Berlin, Heidelberg: Springer-Verlag; 1986. pp. 478-498

[14] Ramamoorthy CV, Li HF. Pipeline architecture. ACM Computing Surveys. 1977;**9**(1):61-102

http://dx.doi.org/10.5772/intechopen.75589

[15] Zuberek WM. Modeling and analysis of simultaneous multithreading. In: Proceedings of the 14th International Conference on Analytical and Stochastic Modeling Techniques and Applications (ASMTA-07), a part of the 21st European Conference on Modeling and Simulation (ECMS'07); Prague, Czech Republic; 2007. pp. 115-120

[16] Suh T, Lee H-HS, Blough DM. Integrating cache coherence protocols for heterogeneous multiprocessor system. Part 2. IEEE Micro. 2004;**24**(5):55-69

[17] Jensen K, Kristensen LM. Coloured Petri Nets—Modeling and Validation of Concurrent Systems. Berlin, Heidelberg: Springer-Verlag; 2009

Chapter 6

Supervisory Control Systems: Theory and Industrial Applications

Hamdi Awad

Additional information is available at the end of the chapter

http://dx.doi.org/10.5772/intechopen.75166

Abstract

Hybrid control system is an exciting field of research where it contains two distinct types of systems: one with continuous dynamics continuous variable dynamic system and the other with discrete dynamics discrete event dynamic system, that interact with each other. The research in the area of hybrid control can be categorized into two areas: one deals with the conventional control systems, and the other deals with the decision making systems. The former addresses the control functions at the low level (field level). The latter addresses the modeling, analysis, and design at the higher level found in the supervision, coordination and management levels. The study of hybrid systems is central in designing intelligent hybrid control systems with high degree of autonomy and it is essential in designing discrete event supervisory controllers for continuous systems.

Keywords: discrete event systems, supervisory control systems, petri nets, embedded systems, industrial processes

1. Introduction

In general automation systems' structure can be categorized into six levels: Sensor/Actuator Level, Machine/Controller Level, Process Automation Level, Operation Unit Level, Plant Level. Trends are making this structure possible and desirable to create streamlined three-level automation systems or even to collapse it [1]. This is due the trend to create embedded control systems, cyber physical systems, networking, and discrete hybrid control systems. The trend to reduce machine size and cost while increasing productivity using nanotechnology, requires new approaches to control systems [2]. Thanks to the increased reliability of industrial PC technology, traditional rack-based PLCs can be replaced with more powerful PC-based control systems. While Industrial PC's provide the highest performance and control capacity, new generations of PC technology based on open embedded operating systems, combine the functions of a PLC and an operator panel in one unit which is applicable to smaller scale applications.

The development of complex man-made systems that perform complicated and interacted tasks, has been accompanied by an ever increasing demand for even more sophisticated modeling and control schemes. The need for a systematic and mathematical approach to analysis, design, and control of complex large scale systems is highly demanded. In fact, In operations research, for example, researchers have been interested for a long time in systematic methods to deal with large-scale systems [3–6]. However, control engineers have taken up this challenge in recent years to develop intelligent models for hybrid dynamical systems [7].

The study of hybrid systems is central in designing intelligent hybrid control systems with high degree of autonomy. Such systems include discrete and continuous activities [8]. The field of discrete vent systems (DES) is relatively a new research area that combines different formalisms, methodologies and tools from Supervisory control theory, artificial intelligence (AI) and operations research (OR) [9]. The domains of DES are: manufacturing automation, communication protocols, robotics, process control, nuclear reactors, space exploration systems, aircraft control systems, fault diagnosis, and refinery systems [5, 10, 11]. Historically, DES were introduced in the early 1980s, in the field of chemical engineering. They quickly gained popularity in modeling and supervision of hybrid systems [12, 13].

The design of supervisory controllers for discrete event systems has received considerable attention in research centers [14–16]. There are methods for designing supervisors based on automata models [17], however, they need exhaustive search over the system states that makes them impractical for systems with large number of states, as the number of states increases the state space explosion problem arises [18, 19]. One way of dealing with these problems is to model discrete event systems with Petri Nets (PNs). In this way the state explosion problem can be avoided. Some recent contributions on the Petri net based supervisory control can be found in [20, 21]. PN models are normally more compact compared with automata-based models and are better suited for the representation of discrete event systems due to its mathematical manipulation and graphical representation [22–24]. Timed Petri nets are common used for industrial control systems [14].

The main objective of this chapter is to explore and step by step construct a supervisory control scheme in the field of DES modeling and control. It also shows how the continuous activities; temperature control, pressure control, etc. are represented by few places resided in the embedded PN models. These objectives can be achieved as follows.

1.1. Investigation of DES modeling and supervision

Types of events that may occur in discrete event systems are controllable, and uncontrollable events (sensors). The latter arises a severe problem when the plant works under control. It cannot be inhibited from firing by the supervisor. Embedded supervisors should be developed to deal with such problems.

http://dx.doi.org/10.5772/intechopen.75166

1.2. Selection of the best modeling formalism

There are many tools to develop DES models e.g. finite state automata, and Petri nets. The latter has a good descriptive power compared with the former. Petri nets models are more compact than automata-based models for representing DES.

1.3. Developing supervisory control scheme

There are two types of supervisors, one is mapping supervisors, and the other is compiled supervisors. The latter has two notable features, its computational demand is small, and its structure can be reconfigured online. The developed supervisors should perform resource allocation, coordination, and deadlock avoidance tasks at the higher level of complex hybrid industrial systems. Ordinary and timed Petri nets are employed in this chapter to structure embedded supervisory control models for industrial processes.

1.4. Testing the proposed scheme

Chemical batch processes and kernel railroad crossing system were employed for testing the proposed control scheme; they have resources scarce, forbidden states, and deadlock problems. This chapter structures embedded supervisory control scheme that deals with the continuous activities at the lower level, as well as the discrete ones at the higher level in efficient manner. The continuous activities can be modeled and supervised using intelligent control schemes.

2. Supervisory control systems

This section gives an introduction to the hybrid nature of complex systems as well as their hierarchical structure. It also includes a brief overview of relative work in the area of DES supervision using finite state automata and Petri nets. A comparison between Petri nets and finite state automata as modeling formalisms for the purpose of supervisor synthesis is given in this section.

2.1. Investigation of DES modeling and supervision

Discrete event systems are useful when dealing with dynamic systems that are not fully modeled by classical models, such as differential or difference equations. It is noted that while differential and difference equation models evolve with time, a discrete event system evolves with the occurrence of events. An comprehensive literature on discrete event systems has appeared in the last 30 years, and their study continues to be an area of ongoing research. DED combines different formalisms, methodologies and tools from control theory (CT), artificial intelligence (AI), and operations research (OR) as shown in **Figure 1** [9]. These methods facilitate the modeling of continuous and discrete activities as a unified approach.

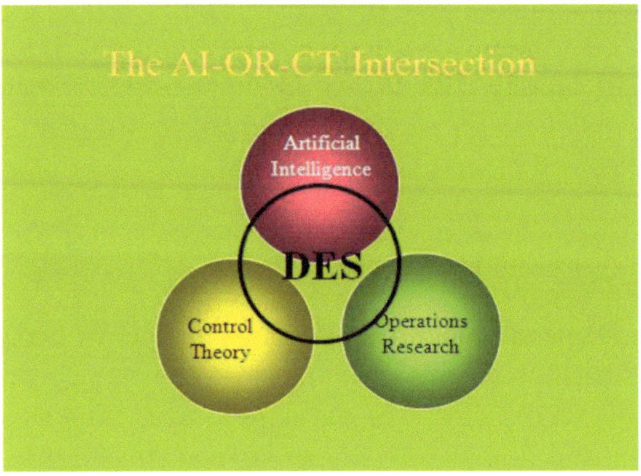

Figure 1. DES: the AI-OR-CT intersection.

System modeling is an important phase in the supervisory system synthesis procedure. There is a set of methods for designing supervisors based on automata system models [16, 17]. The disadvantage of these methods is that they need a huge search over the system states as mentioned in Section 1. One way of dealing with these problems is to model discrete event systems with Petri nets (PNs). Petri net-based solutions have several advantages over finite state automata. These advantages recommends the Petri nets to be used in this chapter. They are listed as follows.

- The states of a Petri net are represented by the possible markings and not by the places. Thus Petri nets give a more compact description.
- The plant and the specifications can be represented graphically in an easily understood format using Petri nets instead of using textual descriptions or mathematical notations, which are difficult to understand.
- Petri net models can be used for the analysis of their properties, performance evaluation and the systematic construction of discrete event supervisors [20, 22].
- The Petri net model allows for the simultaneous occurrence of multiple events.

2.2. Structuring a supervisory control scheme

This subsection shows the use of Petri nets to design a supervisor, including its synthesis techniques, methods of handling uncontrollable and unobservable transitions within the plant structure. The supervision based on place invariants method is employed in this chapter to build an embedded supervisory control systems. In this method, the control objective is to force the process to obey linear constraints in the form of linear inequalities. The ideas of developing such supervisors were borrowed from [21]. The developed supervisory control scheme can be employed to control the processes that have controllable and uncontrollable transitions [25]. Using an embedded Petri net structure, the developed scheme is easy to implement and its computational demand is relatively small. The design method has numerical properties that make it particularly appealing for large scale systems. Mathematically, the developed scheme can be detailed as follows.

A place invariant is an integer vector x that satisfies [21, 25]:

$$x^T\mu = x^T\mu_0 \tag{1}$$

for all reachable marking μ. Thus $x^T\mu$ is constant for all reachable states if x is a place invariant. Place invariants can be computed by finding solutions to Eq. (2).

$$x^T D = 0 \tag{2}$$

Based on the method of place invariants, it is possible to enforce a set of constraints on the plant state μ_p. The plant state is represented by an $n \times 1$ marking vector of non-negative integers, where each vector component is equal to the marking of the corresponding place in the Petri net model of the plant. The supervisory control goal is to restrict the reachable marking vectors of a plant μ_p as:

$$L\mu_p \leq B \tag{3}$$

where L is a $n_c \times n$ integer matrix ($L \in Z^{ncxn}$), and B is a $n_c \times 1$ integer vector ($B \in Z^{nc}$), and n_c is the number of constraints. After adding the slack variables, constraint define in Eq. (3) becomes:

$$L\mu_p + \mu_c \ = B \tag{4}$$

Each place invariant defined in Eq. (6) must satisfy Eq. (2) such that:

$$LD_p + D_c \ = 0 \tag{5}$$

The matrix D_c contains the arcs that connect the controller places to the transitions of the process net. So, given the Petri net model of the process D_p and the constraints, the Petri net controller D_c is constructed see [25].

$$D_C = -LD_P \tag{6}$$

If the initial marking defined in Eq. (7) does not violate the given set of constraints, these constraints can be enforced by a supervisor with the incidence matrix D_c [21].

$$\mu_{C0} = B - L\mu_{p0} \tag{7}$$

where μ_{p0} is the $n \times 1$ initial plant marking vector of non-negative integers. The supervisor is a Petri net with incidence matrix D_c made up of the process net's transitions and a separate set of places. With the addition of supervisor places the overall system is given by.

$$D = \begin{bmatrix} D_p \\ D_C \end{bmatrix} \mu = \begin{bmatrix} \mu_p \\ \mu_C \end{bmatrix} \tag{8}$$

This method admits the structure of the process net as well as a set of specifications. This is because the constraints on these transitions is a subset of the specifications. Supervisors are used to insure that the behavior of the plant does not violate a set of constraints defined in Eq. (3) under a variety of operating conditions. Every single constraint is transformed to a marking invariant that corresponds to a place invariant of the supervised system. The regulatory actions of the supervisor are based on observations of the plant state, resulting in feedback control. More details about the P-invariant-based supervisory control schemes can be found in [17, 21, 25].

2.3. Step by step developing P-invariant-based supervisory control scheme

The most common examples of hybrid systems are batch processes which are characterized by combination of discrete and continuous dynamics [17, 26]. Batch processes are currently used in the chemical and food process industries. A comprehensive model of batch systems has to include discrete event aspects as well as continuous ones. As a consequence, their automation and optimization pose difficult issues mainly because it is necessary to operate concurrently with continuous and discrete models.

Batch plants consist of many transport resources (*transporters*) like valves and pipes, and processing resources (*processors*) like mixing tanks, batch reactor vessels, and other container like units [27–29]. Transporters and processors are involved in transforming a batch from raw materials to final products.

The main problems inherent to the modeling and supervision of batch processes are:

- *The hybrid nature of the process.* State variables like the tank level or the pump speed are continuous, others like the on-off valves, are discrete-state components. Moreover, the whole process behaves as a cycle of discrete events.
- *The variety of knowledge.* Some elements of the process can be described by physical equations, e.g. the tank level, and others by experimental models, e.g. the behavior of the pump described by a transfer function.

The event driven part of a batch plant is modeled using Petri nets. The most appropriate method in batch plants modeling is the Petri net Bottom-up synthesis method [30, 31], where plants in process industries generally exhibit less flexible structure than manufacturing processes and the individual process units are well defined and standardized [29].

This section uses the bottom-up approach to Petri net modeling of the chemical batch processes. It also employs the P-invariant supervisory control scheme described in Section 2.2 to structure an embedded PN-supervisory control model of the batch process.

2.3.1. *Example 1: a simple batch plant*

This process used to illustrate the idea of supervisor design for the purpose of resource allocation for two process lines chemical process shown in **Figure 2**.

In this figure, two mixing tanks shared the same supply tank and only one tank can be filled at a time. Based on the receipt given in **Table 1**, using Petri net tool ver. 2.1 [33] the PN model of the two individual process lines shown in **Figure 2** is constructed as depicted in **Figure 3**. This model is structured using the bottom-up synthesis method [31].

In this sample, the places P_{m3} are corresponding to the outlet valve of the supply tank. The invariant-based supervision discussed in Section 2.2 is employed for the purpose of supervisor

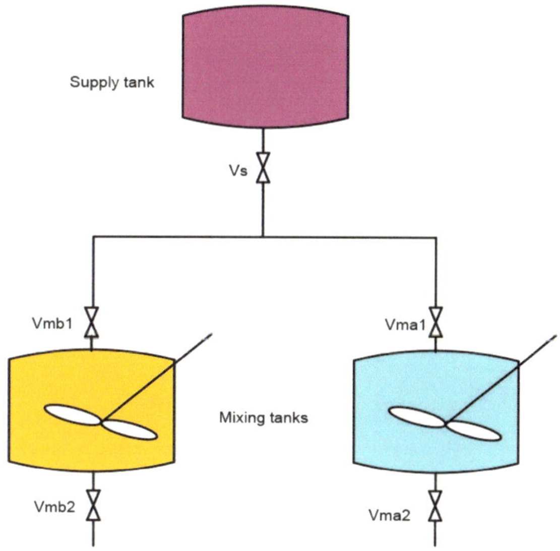

Figure 2. Batch process cell.

Place	Associated action
P_{m1}	Process ready
P_{m2}	Open the inlet valve of the mixing tank
P_{m3}	Open the valve of supply tank
P_{m4}	Stir the content of mixing tank
P_{m5}	Discharge the mixing tank (open the outlet valve)
Transition	Associated event
T_{m1}	Start a new batch
T_{m2}	Mixing tank is filled
T_{m3}	Duration of mixing operation is vanished
T_{m4}	Empty the mixing tank

Table 1. Places and transitions for each submodel of **Figure 3**.

synthesis. This is particularly interesting, because the resulting supervisory mechanism is computed efficiently. All transitions are assumed to be controllable.

It is clear that, if transitions T_{ma1} and T_{mb1} are fired, the place P_{m3} contains two tokens and therefore the Petri net model is not safe. The safeness of the place P_{m3} is required, because it represents the operation of opening and closing the outlet valve of the supply tank. The double-booking problem should be avoided in this case, otherwise, the situation is considered as a malfunction. To overcome this problem, the supervisor has to be designed to co-ordinate the two mixers in such a way that only one will be filled at a time. This requirement is written as:

$$\mu_{m3} \leq 1 \tag{9}$$

where μ_3 is the marking vector component that corresponds to place P_{m3}. The requirement can be easily transformed to the form Eq. (6) with the plant marking vector being:

$$\mu_p = \begin{bmatrix}\mu_{ma1} & \mu_{ma2} & \mu_{m3} & \mu_{ma4} & \mu_{ma5} & \mu_{mb1} & \mu_{mb2} & \mu_{mb4} & \mu_{mb5}\end{bmatrix}^T$$

$\mu_{p0} = [1 \ \ 0 \ \ 0 \ \ 0 \ \ 0 \ \ 1 \ \ 0 \ \ 0 \ \ 0]^T, L = [0 \ \ 0 \ \ 1 \ \ 0 \ \ 0 \ \ 0 \ \ 0 \ \ 0 \ \ 0]$ and $B = 1$.

Given the incident matrix of the PN model shown in **Figure 3**; D_p:

$$D_p = \begin{bmatrix} -1 & 0 & 0 & 1 & 0 & 0 & 0 & 0 \\ 1 & -1 & 0 & 0 & 0 & 0 & 0 & 0 \\ 1 & -1 & 0 & 0 & 1 & -1 & 0 & 0 \\ 0 & 1 & -1 & 0 & 0 & 0 & 0 & 0 \\ 0 & 0 & 1 & -1 & 0 & 0 & 0 & 0 \\ 0 & 0 & 0 & 0 & -1 & 0 & 0 & 1 \\ 0 & 0 & 0 & 0 & 1 & -1 & 0 & 0 \\ 0 & 0 & 0 & 0 & 0 & 1 & -1 & 0 \\ 0 & 0 & 0 & 0 & 0 & 0 & 1 & -1 \end{bmatrix}$$

The supervisor can be computed by Eqs. (6) and (7) as follows:

$D_c = -LD_p = [-1 \ \ 1 \ \ 0 \ \ 0 \ \ -1 \ \ 1 \ \ 0 \ \ 0]$, $\mu_{C0} = B - L\mu_{p0} = 1 - 0 = 1$.

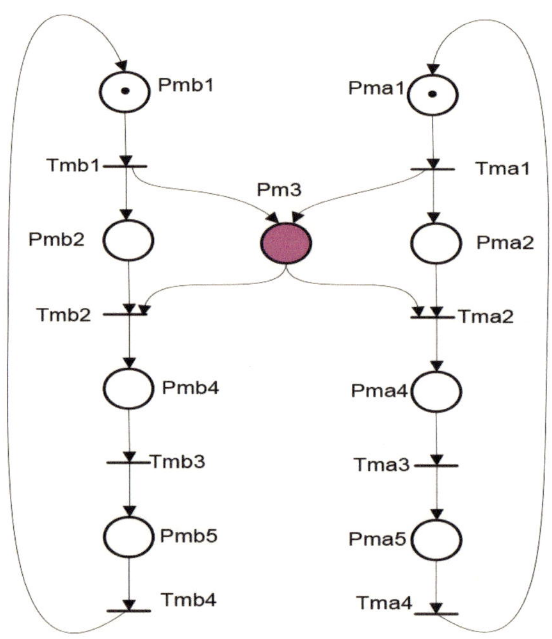

Figure 3. Petri net model of the overall system.

The supervisor consists of a single place that is connected to the plant Petri net as shown in **Figure 4**. The marking invariant that is enforced by the supervisor is:

$$\mu_{m3} + \mu_{c1} = 1 \tag{10}$$

The simulation of the unsupervised system via Reachability graph analysis method of PN indicates that the unsupervised plant has 16 reachable states, one of them indicates the absence of safety condition, this marking vector is: $\mu_4 = [0\,1\,2\,0\,0\,0\,1\,0\,0]^T$ i.e. $\mu_{m3} = 2$. On the other hand, the simulation of the supervised system indicates that the supervised plant has 15 reachable states. In this case, the supervisor eliminates the marking μ_4 which is forbidden state, and all the reachable states satisfy the safety condition. This procedure can be generalized for more complex batch processes such as coordination, deadlock avoidance, and resource allocation discussed in [17].

Quiz 1: With the help of our work in [14], can you model the chemical batch process shown in **Figure 2** using timed Petri nets?

2.3.2. *Control of continuous activities as a part of hybrid systems*

The main objective of this subsection is to control the continuous part of a complex batch process shown in **Figure 5**. This process comprises six input buffers, two mixing tanks and two reactor vessels. In this case, heating and cooling are continuous variables of this batch process. The preparation of the input substances takes place in two mixing tanks to which the raw materials are supplied from three supply tanks (buffers). The substance is composed from one of the two basic components (component 'a' or component 'b') that is diluted to the required concentration by component 'c'. The filling of the mixing tank is controlled by the on/off valve Vma in

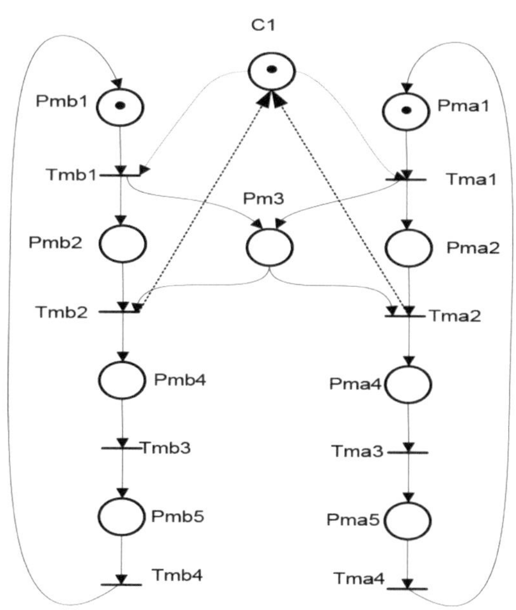

Figure 4. Petri net model of the supervised system.

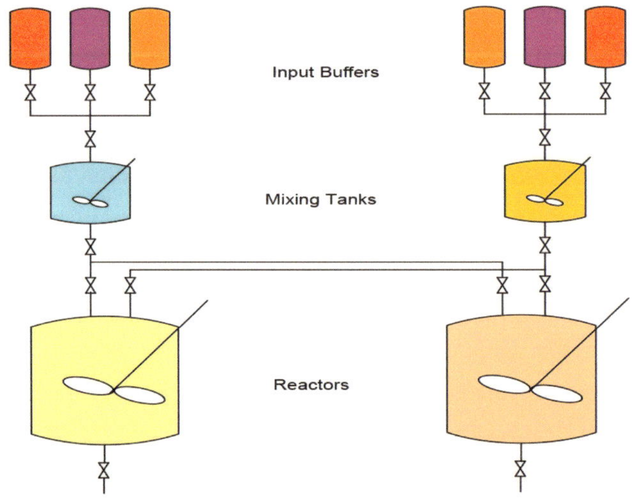

Figure 5. Batch process cell with two mixers and two reactors.

combination by one of the supply tank valves Vsa, Vsb, or Vsc. The discharging of the mixer is controlled by the on/off valve Vmb. The level in the mixing tank is measured by the level sensor; L. The required quantities of each input depend on the recipe detailed in [17].

The mathematical differential equations of the heat system of the reactor is detailed in [22]. In this case, the temperature of the reaction for each reactor is controlled using fuzzy neural systems-based local controllers [32] that are designed for each of heating and cooling phases in each process line. It is controlled by feeding hot or cold glycol through the reactor jacket, which surrounds the reactor vessel.

Using the P-invariant supervisory control method discussed in Section 2.2, the embedded PN-based supervisory control model is structured using Petri net tool ver. 2.1 [33]. A part of this embedded model is shown in **Figure 6**. The main objective of this subsection is to show how can activate the continuous activities resided inside the embedded model of the process. It also shows the control result of heating phase as a set of continuous places resided in the embedded model using fuzzy neural controller. The full simulation results can be found in our work [17]. To show the firing of the continuous activity resided in the embedded model, consider the initial marking vector, μ_0 defined below.

$$\mu_0 = [000000000000000000001000000000000000000\ 00111111]^T$$

Starting from this initial marking vector, the simulation of a part of the embedded PN-based model shown in **Figure 6** [17], can be carried out. When the transition T_{rb3} is fired, the fuzzy neural controller resided in place 8; Prb8, is activated as shown in **Figure 6** and the marking vector becomes, μ_{HR}.

$$\mu_{HR} = [10000000000000000000001000000000001011000001111]^T$$

The response of the reaction at the desired set point 50°C is depicted in **Figure 7**.

http://dx.doi.org/10.5772/intechopen.75166

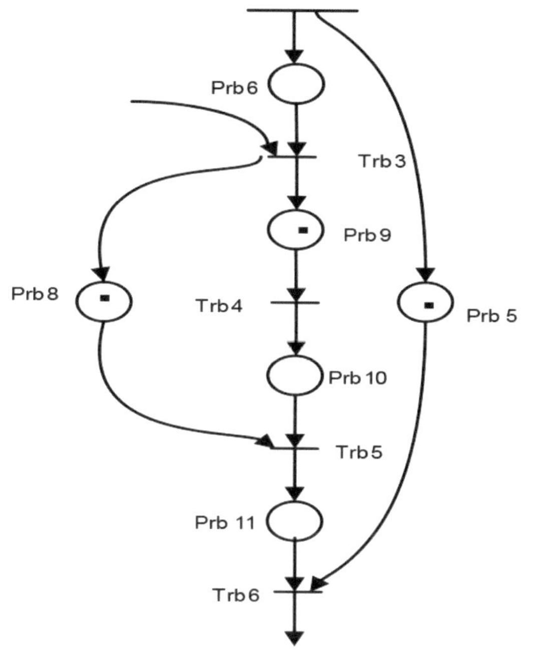

Figure 6. A part of the embedded PN model of batch process shown in **Figure 5**.

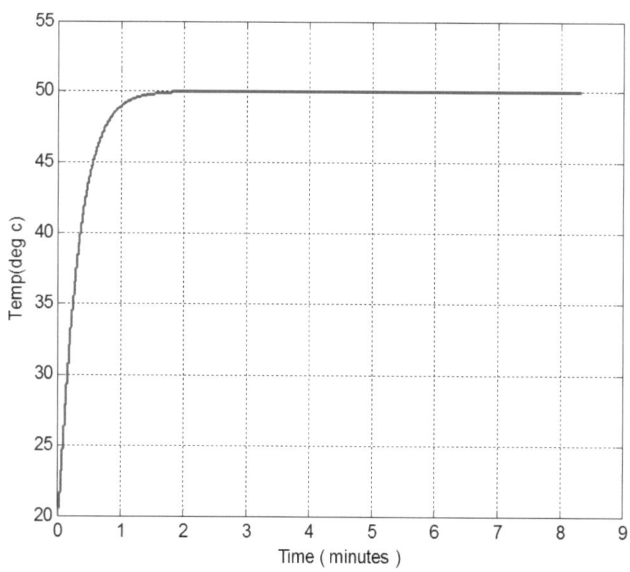

Figure 7. Reactor temperature for the heating phase.

2.3.3. *Example 3: KRC modeling and supervision using timed PNs*

The kernel railroad crossing (KRC) shown in **Figure 8** is a standard benchmark in real time systems [14]. When a train is sensed to approach the crossing, a signal is sent to the supervisor that sends a command to the specified gate that is closed to prevent cars crossing that survives ourselves. Having more than one track and more than one train may enter crossing zone, leads to a complicated situation that is out of our paper scope. For simplicity let us merge the two depicted zones as one zone (region). The Petri net of the system is depicted in **Figure 9**. The train needs one time unit (t.u.) to enter the R-Y zoon and five (t.u.) to leave it for departure phase. The gate needs no time to start closing and requires two (t.u.) to be completely closed. It needs extra two (t.u.) to be completely opened after firing the transition T_5. The problem arises

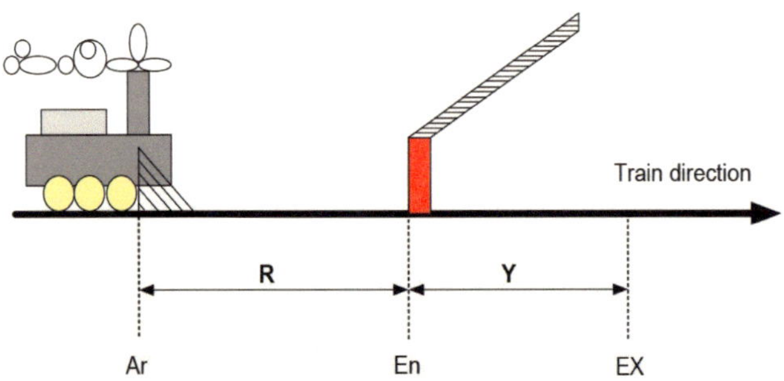

Figure 8. A kernel railroad crossing system.

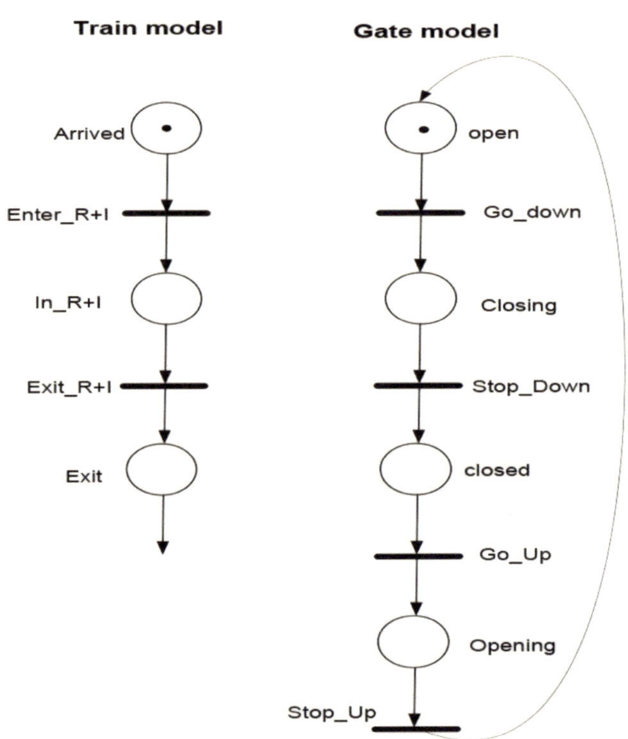

Figure 9. The simplified Petri net model of **Figure 8**.

at the beginning of opening the gate that has unknown time units "[0, ∞]". This is due to the reason that no expectation for beginning the departure phase of the train; its departure depends on the passenger riding. The analysis should be performed to get the exact period time unit required to activate the transition T_5 every departure phase. This means that this transition is continuously evaluated. The system comprises two tasks, train task and gate task. In our work [14], consider that the controllable events are the beginning of each task. However, the accomplishing of the tasks is uncontrollable. Therefore, our goal was to control the beginning of the tasks in order to obtain safe arrival and departure of the train. Also, the synthesized control scheme should avoid the forbidden state (P_2, P_4). This means that the train in the R-Y zone and the gate still opened.

Using Petri net tool software ver. 2.1 [33], the system incidence matrix is of the PN model shown in **Figure 9** is:

$$D_P = \begin{bmatrix} -1 & 0 & 0 & 0 & 0 & 0 \\ 1 & -1 & 0 & 0 & 0 & 0 \\ 0 & 1 & 0 & 0 & 0 & 0 \\ 0 & 0 & -1 & 0 & 0 & 1 \\ 0 & 0 & 1 & -1 & 0 & 0 \\ 0 & 0 & 0 & 1 & -1 & 0 \\ 0 & 0 & 0 & 0 & 1 & -1 \end{bmatrix}$$

In this simulation, there are 11 reachable states starting from the initial marking vector μ_0 to the final vector μ_{11}. The firing sequence shows that the marking vector, $\mu_1 = [0\,1\,0\,1\,0\,0\,0]^T$ is a forbidden state. It is clear that the marking vector μ_1 includes the forbidden state (P_2, P_4). Based on P-invariant, a supervisory control scheme for the KRC system is synthesized in Section 2.2. The constraints vector is $L = [0 \quad 1 \quad 0 \quad 1 \quad 0 \quad 0 \quad 0]$, the controller incidence matrix is $D_c = -LD_p = [-1 \quad 1 \quad 1 \quad 0 \quad 0 \quad -1]$, the initial marking of the controller is $\mu_{c0} = B - L\mu_0 = 0$, and $B = 1$. The developed supervised time Petri net of the KRC system is depicted in **Figure 10**. There are 10 reachable states starting from the initial marking vector μ_0 to the final vector μ_{10} and the supervisor eliminates the forbidden state vector $\mu_1 = [0\,1\,0\,1\,0\,0\,0]^T$.

Another issue, in distributed hybrid systems, each process line is controlled by its own logic controller and supervisory part resides in another level. The interaction among different modules is performed through synchronized transitions. In practical implementation, it is difficult to achieve such synchronization among the logic controllers of the embedded systems. Because of the communication delays, it cannot be guaranteed that transitions in different controllers fire simultaneously. One way for dealing with this problem is to define the firing order of the transitions. In the cases when the two logic controllers share the same resource and the supervisor performs the resource allocation such as indicated in **Figure 4**, the transition that reserve the resource must be fired in the supervisor first and then in the local controller. In the opposite case the communication delay would allow double booking problem of the shared resource by the two controllers or even deadlock. Due to the pages limitation, more details about this implementation problem the readers can be directed to read our work detailed in [34].

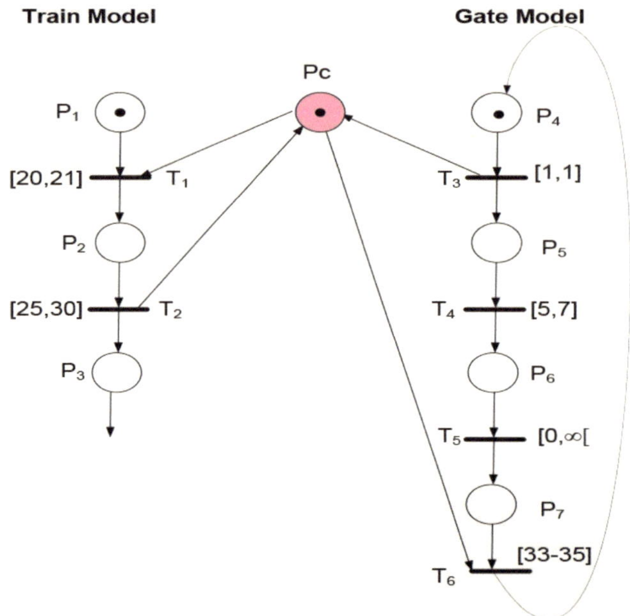

Figure 10. The supervised timed Petri net of the KRC (**Figure 8**).

Quiz 2: With the help of our work in [17], can the reader complete the missing part of the supervised embedded PN model depicted in **Figure 6** using P-invariant supervisory control discussed in Section 2.2?

Quiz 3: With the help of our work in [34], can you overcome the communication problems through the embedded PN model of Quiz 2?

3. Conclusions

Hybrid systems modeling and supervision have been used extensively in automation, robotics, and manufacturing applications. Different frameworks for dynamic supervisory controllers are used in flexible manufacturing systems and automated batch processes. The high-level system changes in hybrid systems are modeled as discrete event dynamic systems, while the low-level systems changes are modeled as continuous variable dynamic systems. The major issue in studying hybrid systems is the consistency between continuous and discrete models evolution. Petri nets possess many assets as models for DES. They provide more compact representation for larger reachable state spaces, and increase the behavioral complexity compared with automata-based models. Batch plants are common examples of hybrid systems. In this chapter Petri net embedded models were developed by abstracting the behavior of hybrid systems. As a final conclusion, the work in this chapter allows the readers to design and analysis their own supervisory control schemes using Petri net tools ver. 2.1 or higher [33]. Although the developed schemes are tested using batch chemical processes, they are promising to control complex industrial automated processes. This chapter opens several research directions to be considered and investigated. It may extend this work to optimization of supervisory control schemes, modeling and supervision of hybrid industrial systems using timed Petri nets, and implementing the proposed schemes for large scale systems.

Author details

Hamdi Awad[1,2]

Address all correspondence to: awadhaa@yahoo.co.uk

1 College of Engineering, Shaqra University, Dawadmi, KSA

2 Faculty of Electronic Engineering, Menoufia University, Egypt

References

[1] Strasser T, Froschauer R. Autonomous application recovery in distributed intelligent automation and control systems. Intelligent Automation and Control Systems, IEEE Transactions on Systems Man and Cybernetics Part C (Applications and Reviews). 2012;**42**:1054-1070. DOI: 10.1109/TSMCC.2012.2185928

[2] Lu K. Nanoparticulate Materials: Synthesis, Characterization, and Processing. USA: John Wiley & Sons, Inc.; 2013. p. 464, ISBN: 978-1-118-29142-9. DOI: 10.1002/9781118408995

[3] Lemmon MD, He KX, Markovsky I. A tutorial introduction to supervisory hybrid systems. Technical Report of the ISIS Group at the University of Notre Dame ISIS-98-004; 1998

[4] Antsaklis PJ. Intelligent control. In: Encyclopedia of Electrical and Electronics Engineering. Vol. 10. USA: John Wiley & Sons, Inc.; 1999. pp. 493-503. DOI: 10.1002/047134608X.W1019

[5] Böhme TJ, Frank B. Hybrid Systems, Optimal Control and Hybrid Vehicles: Theory, Methods and Applications. Berlin: Springer Nature; 2017, 519 p. DOI: 10.1007/978-3-319-51317-1

[6] Jalivand A, Khanmohammadi S. Integrating of event detection and mode recognition in hybrid systems by fuzzy Petri nets. In: Proceeding of the IEEE Conference on Robotic, Automation, and Mechatronics; 1–3 December 2004; Singapore. pp. 265-270

[7] Giua A, Silva M. Modeling, analysis and control of discrete event systems: A Petri net perspective. IFAC-PapersOnLine. 2017;**50**:1772-1783. DOI: 10.1016/j.ifacol.2017.08.156

[8] Chen CH, Dai JH. Design and high-level Synthesis of hybrid controller. In: Proceedings of the IEEE International Conference on Networking, Sensing ,and Control; 21–23 March 2004; Taipei, Taiwan. pp. 433-438. DOI: 10.1109/ICNSC.2004.1297477

[9] Ho Y-C. A road map for DEDS research. IFAC Proceedings Volumes. 1993;**26**:663-669. DOI: 10.1016/S1474-6670(17)49211-1

[10] Villani E, Miyagi PE, Valette R. Landing system verification based on Petri nets and a hybrid approach. IEEE Transactions on Aerospace and Electronic Systems. 2006;**42**:1420-1436. DOI: 10.1109/TAES.2006.314582

[11] Sobh TM, Benhabib B. Discrete event and hybrid systems in robotics and automation: An overview. IEEE Robotics and Automation Magazine. 1997;**4**:16-19. DOI: 10.1109/100.591642

[12] Cohn G, Cnet PM, Quadrat JP. Linear system theory for discrete event systems. In: Proceedings of the 23rd IEEE Conference on Decision and Control, Las Vegas; 12–14 December 1984; USA. p. C1. DOI: 10.1109/CDC.1984.272058

[13] Silva M, Recalde L. On fluidification of Petri nets: From discrete hybrid and continuous models. Annual Reviews in Control. 2004;**28**:253-266. DOI: 10.1016/j.arcontrol.2004.05.002

[14] Allaa H, Awad H, Anwar A, El-Hajri E. Modelling and control of KRC systems using TPN and TA. International Journal of Advanced Research in Electrical, Electronics and Instrumentation Engineering, USA. 2015;**4**: ISSN (Online): 2278-8875:1111-1130

[15] Hopcroft JE, Ullman JD. Introduction to Automata Theory, Languages, and Computation. 3rd ed.. Addison-Wesley series in Computer Science, ISBN 0-201-02988-X, 530 p ed1979

[16] Ramadge PJ, Wonham WM. The control of discrete event systems. Proceedings of the IEEE. 1989;**77**:81-97. DOI: 10.1109/5.21072

[17] Anwar A. Modelling and supervision of discrete event systems [thesis]. Egypt: Menoufia University; 2007

[18] Giua A, DiCesare F. Petri net structural analysis for supervisory control. IEEE Transactions on Robotics and Automation. 1994;**10**:185-195. DOI: 10.1109/70.282543

[19] Music G, Matko D. Petri net based control of a modular production system. In: Proceedings of the IEEE International Symposium on Industrial Electronics, Vol. 3; 12–16 July 1999; Bled, Slovenia. pp. 1383-1388. DOI: 10.1109/ISIE.1999.796909

[20] Wang S, Wang C, Zhou M. Design of optimal monitor-based supervisors for a class of Petri nets with uncontrollable transitions. IEEE Transactions on Systems, Man, and Cybernetics: Systems. 2013;**43**:1248-1255. DOI: 10.1109/TSMC.2012.2235427

[21] Iordache MV, Antsaklis PJ. Supervision based on place invariants: A survey. Technical Report of the ISIS Group at the University of Notre Dame ISIS-2004-003; July, 2004

[22] Murata T. Petri nets: Properties, analysis and applications. Proceedings of the IEEE. 1989; **77**:541-580. DOI: 10.1109/5.24143

[23] David R, Alla H. Petri nets for modeling of dynamic systems: A survey. Automatica. 1994; **30**:175-202. DOI: 10.1016/0005-1098(94)90024-8

[24] Cheung KS, Cheung TY, Chow KO. A Petri-net-based synthesis methodology for se-case-driven system design. Journal of Systems and Software. 2006;**79**:772-790. DOI: 10.1016/j.jss.2005.06.018

[25] Moody JO, Antsaklis PJ. Petri net supervisors for DES with uncontrollable and unobservable transitions. IEEE Transactions on Automatic Control. 2000;**45**:462-476. DOI: 10.1109/9.847725

[26] Lennartson B, Egardt B, Tittu M. Hybrid systems in process control. In: Proceedings of 33rd Conference on Decision and Control; 14–16 December 1994; Lake Buena, USA. IEEE. pp. 3587-3592. DOI. 10.1109/CDC.1994.411706

[27] Tittus M, Lennartson B. Hierarchical supervisory control for batch processes. IEEE Transactions on Control Systems Technology. 1999;**7**:542-554. DOI: 10.1109/87.784418

[28] Falkman P, Lennartson B, Tittus M. Specification of a batch plant using process algebra and Petri nets. In: Proceedings of the IEEE International Conference on Automation Science and Engineering; 1–2 August 2005; Edmonton, Canada. pp. 339-344. DOI: 10.1109/COASE.2005.1506792

[29] Music G, Matko D. Petri net based supervisory control of flexible batch plant. IFAC Proceedings Volumes. 15–17 July 1998;**31**:941-946. DOI: 10.1016/S1474-6670(17)41919-7

[30] Luo J, Zhou M. Petri-net controller synthesis for partially controllable and observable discrete event systems. IEEE Transactions on Automatic Control. 2017;**62**:1301-1313. DOI: 10.1109/TAC.2016.2586604

[31] Wang S, You D, Wang C. Optimal supervisor synthesis for Petri nets with uncontrollable transitions: A bottom-up algorithm. Information Sciences. 2016;**363**:261-273. DOI: 10.1016/j.ins.2015.11.003

[32] Awad H. Fuzzy neural networks for modelling and controlling dynamic Systems [thesis]. Wales, England: Cardiff University; 2001

[33] Petri Net Toolbox for Matlab. Department of Automatic Control and Industrial Informatics of the Technical University "Gh. Asachi" of Iasi, Romania. Available from: http://www.ac.tuiasi.ro/pntool/ [Accessed: May 2007]

[34] Gomaa M, Awad H, Anwar A. Design and implementation of supervisory control schemes in industrial automation systems. In: Proceedings of the IEEE International Conference of Computer Engineering & Systems, ICCES'08; 25–27 November 2008; Cairo, Egypt. pp. 398-404. DOI: 10.1109/ICCES.2008.4773035

Chapter 7

Process Petri Nets with Time Stamps and Their Using in Project Management

Ivo Martiník

Additional information is available at the end of the chapter

http://dx.doi.org/10.5772/intechopen.76769

Abstract

Process Petri nets with time stamps (PPNTS) are the newly introduced class of low-level Petri nets, whose definition and the properties are the main topic of this chapter; they generalize the properties of Petri net processes in the area of design, modeling and verification of generally parallel systems with the discrete time. Property-preserving Petri net process algebras (PPPAs) were originally designed for the specification and verification of manufacturing systems. PPPA does not need to verify composition of Petri net processes because all their algebraic operators preserve the specified set of the properties. These original PPPAs are generalized for the class of the PPNTSs in this chapter. The new COMP, SYNC and JOIN algebraic operators are defined for the class of PPNTS and their chosen properties are proved. With the support of these operators, the PPNTSs can be extended also to the areas of project management and the determination of the project critical path with the support of the critical path method (CPM). The new CPNET subclass of PPNTS class is defined in this chapter. It is specially designed for the generalization of the CPM activity charts and their properties. This fact is then demonstrated on the simple project example and its critical path and other property specifications.

Keywords: process Petri nets with time stamps, property-preserving Petri net process algebras, critical path method, discrete time, property preservation, parallel systems modeling

1. Introduction

There are currently a number of formally defined classes of **Petri nets** [1, 2] available for modeling of generally parallel systems. When studying distributed parallel programming systems, real-time systems, economic systems and many other types of systems, it plays a role modeling of the time variables associated with individual system events, the duration of the studied activities, the time history of the modeled system and many other time characteristics. Special classes of Petri nets were introduced for the modeling of these types of systems with discrete time and their properties were studied in detail. **Time Petri nets** and **timed Petri nets** [3, 4] are currently the two most important classes of low-level Petri nets that use the concept of discrete time in their definition. Other classes of low-level Petri nets with discrete time are introduced and discussed for instance in [5–7]. It can be stated that most of the currently studied classes of Petri nets with discrete time use only the relative time variables usually related to the specific marking of the given Petri net. This fact can then cause difficulties, for example, in modeling complex time-synchronized distributed systems in which an external time source is usually available and individual components of this system must be synchronized with this external time source.

Process Petri nets (PPN) [8] were primarily introduced as a special subclass of classic low-level Petri nets for their using in the area of workflow management. PPN is a continuous Petri net that include within the set of all its places the unique input place, the unique output place and a finite set of so-called resource places which may contain, along with the input place, the tokens in the entry marking of the given PPN. These tokens located in the entry marking of PPN at the resource places usually represent the permanent resources of the modeled system. The given PPN can pass into its exit marking that is reachable from its entry marking by performing the final sequence of the transition firings. The tokens of the PPN's exit marking may be then located only at its single output place and also at its resource places.

Process Petri nets with time stamps (PPNTS) are the newly introduced class of low-level Petri nets whose definition and the properties are the main topics of this chapter. PPNTS generalize the properties of PPNs in the area of design, modeling and verification of generally parallel systems with the discrete time.

Property-preserving Petri net process algebras (PPPA) [9] were originally designed for the specification and verification of manufacturing systems. PPPA has four types of operators: extensions, compositions, refinements and reductions. All operators can preserve about 20 PPN's properties (some of them under additional conditions), such as liveness, boundedness, reversibility, RC-property, traps, siphons, proper termination, and so on. PPPA does not need to verify composition of PPNs because all their algebraic operators preserve the specified set of the properties. Hence, if the source PPNs satisfy the desirable properties, each of the composite PPN, including the PPN that models the resulting system itself, also satisfies these properties. These original PPPA are generalized for the class of the PPNTS in this chapter and their properties of proper-formed, well-formed and pure-formed PPNTS are then newly introduced. The new **COMP**, **SYNC** and **JOIN** algebraic operators are defined for the class of PPNTS and their chosen properties are proved.

With the support of these operators, the PPNTS can be extended also to the areas of the project management and the determination of the project critical path with the support of the critical path method (CPM) [10]. The new CPPNET subclass of PPNTS class is then defined in this chapter to represent pure-formed time-dependent processes. It is specially designed for the generalization of the CPM activities charts and their properties. This fact is then demonstrated on the simple project example and its critical path and other properties specification.

This chapter is arranged into the following sections: Section 2 explains the base term of this chapter, that is, process Petri nets with time stamps and introduces the terms of proper-formed, well-formed and pure-formed PPNTS; Section 3 discusses algebraic operators **COMP** and **SYNC** defined over the class of PPNTS and their main properties; Section 4 then introduces the special subclass CPPNET of the PPNTS class and explains its use in the area of the project management to represent pure-formed time-dependent processes with using of PPNTSs and to find critical paths for these processes similarly as in the case of the well-known critical path method (CPM). Finally, Section 5 gives the conclusions of the research to conclude the chapter.

2. Process Petri nets with time stamps and their properties

Let $\mathbf{N}$ denote the set of all natural numbers, $\mathbf{N} := \{1, 2, \ldots\}$; $\mathbf{N}_0$ the set of all non-negative integer numbers, $\mathbf{N}_0 := \{0, 1, 2, \ldots\}$; $\varnothing$ the empty set; $|A|$ the cardinality of the given set A; $\mathscr{P}(A)$ denotes the family of all the subsets of the given set A; $f\colon A \to B$ a function on a domain A to a codomain B; $\neg$ the logical negation operator. Let $(A \subset \mathbf{N}_0) \wedge (\exists n \in \mathbf{N}\colon |A| = n) \wedge (A \neq \varnothing)$; then $max(A) := x$, where $(x \in A) \wedge (\forall y \in A\colon x \geq y)$. **Multiset** M over a nonempty set S is a function $M\colon S \to \mathbf{N}_0$. The non-negative number $M(a) \in \mathbf{N}_0$, where $a \in S$, denotes the number of occurrences of the element a in the multiset M. The multiset M over a nonempty set S will be represented by the notation $M := [a^{M(a)}, b^{M(b)}, c^{M(c)}, \ldots] = [a, \ldots, a, b, \ldots, b, c, \ldots, c, \ldots]$, where $S := \{a, b, c, \ldots\}$. Notation S_{MS} then denotes the class of all the multisets over the set S.

Definition 1. Let A be a nonempty set. By the (nonempty finite) **sequence** σ over the set A we understand a function $\sigma\colon \{1, 2, \ldots, n\} \to A$, where $n \in \mathbf{N}$. Function $\varepsilon\colon \varnothing \to A$ is called the **empty sequence** on the set A. We usually represent the sequence $\sigma\colon \{1, 2, \ldots, n\} \to A$ by the notation $\sigma = \langle a_1, a_2, \ldots, a_n \rangle$ of the elements of the set A, where $a_i = \sigma(i)$ for $1 \leq i \leq n$. Empty sequence $\varepsilon\colon \varnothing \to A$ on the set A we usually represent by the notation $\varepsilon = \langle\rangle$. We denote the set of all finite (and possible empty) sequences over the set A by the notation A_{SQ}.

If $\sigma = \langle a_1, a_2, \ldots, a_n \rangle$ and $\tau = \langle b_1, b_2, \ldots, b_m \rangle$ are the finite sequences, where $\sigma \in A_{SQ}$, $\tau \in A_{SQ}$, $n \in \mathbf{N}$, $m \in \mathbf{N}$, then by the **concatenation of the sequences** σ and τ, denoted by $\sigma{+}{+}\tau$, we understand the finite sequence $\sigma{+}{+}\tau := \langle a_1, a_2, \ldots, a_n, b_1, b_2, \ldots, b_m\rangle$. The following functions are defined:

i. $length\colon A_{SQ} \to \mathbf{N}_0$, so that: $length(\sigma) := n$, $length(\varepsilon) := 0$,

ii. $elements\colon A_{SQ} \to \mathscr{P}(A)$, so that: $elements(\sigma) := \{a \mid \exists i, 1 \leq i \leq n\colon a = \sigma(i)\}$, $elements(\varepsilon) := \varnothing$,

iii. $prefix\colon A_{SQ} \times \mathbf{N}_0 \to A_{SQ}$, so that:

$prefix(\langle a_1, a_2, \ldots, a_n\rangle, m) := \langle a_1, a_2, \ldots, a_m\rangle$, if $m \leq n$,

$prefix(\langle a_1, a_2, \ldots, a_n\rangle, m) := \langle a_1, a_2, \ldots, a_n\rangle$, if $m > n$,

$prefix(\varepsilon, m) := \varepsilon$,

iv. *suffix*: $A_{SQ} \times \mathbf{N}_0 \rightarrow A_{SQ}$, so that:

$suffix(<a1, a2, \ldots, a_n>, m) := < a_{m+1}, a_{m+2}, \ldots, a_n>, \text{ if } m < n,$

$suffix(<a1, a2, \ldots, a_n>, m) := \varepsilon, \text{ if } m \geq n,$

$suffix(\varepsilon, m) := \varepsilon,$

v. *create*: $\mathbf{N} \times A \rightarrow A_{SQ}$, so that: $create(n, a) := < a, a, \ldots, a>$, where $length(<a, a, \ldots, a>) = n$,

vi. *sort*: $(\mathbf{N}_0)_{SQ} \rightarrow (\mathbf{N}_0)_{SQ}$, so that: $sort(\sigma) := \rho$,

where $(\rho = <b_1, b_2, \ldots, b_n>) \wedge (b_1 \leq b_2 \leq \ldots \leq b_n) \wedge ([a_1, a_2, \ldots, a_n] = [b_1, b_2, \ldots, b_n])$.

We use the following subsets of the set $(\mathbf{N}_0)_{SQ}$:

- $\mathbf{N}^\# := \{\sigma \in (\mathbf{N}_0)_{SQ} \mid (\sigma = \varepsilon) \vee ((\sigma = <a_1, a_2, \ldots, a_n>) \wedge (a_1 \leq a_2 \leq \ldots \leq a_n)), n \in \mathbf{N}\}$,
- $\mathbf{N}^0 := \{\sigma \in (\mathbf{N}_0)_{SQ} \mid (\sigma = \varepsilon) \vee ((\sigma = <0, 0, \ldots, 0>) \wedge (length(\sigma) = n)), n \in \mathbf{N}\}$.

Thus, the elements of the set $\mathbf{N}^\#$ constitute the empty sequence ε and all the finite ascending ordered sequences σ consisting of non-negative integer numbers. Similarly, the elements of the set $\mathbf{N}^0$ then form an empty sequence ε and all the sequences in the form <0, 0, …, 0 > of any finite length.

Definition 2. Net *NET* is an ordered triple *NET* := (*P*, *T*, *A*), where *P* is finite nonempty set of **places**, *T* is finite set of **transitions**, $P \cap T = \varnothing$, and *A* is finite set of **arcs**, $A \subseteq (P \times T) \cup (T \times P)$. □

The given net *NET* is then described with a bipartite graph containing a finite nonempty set *P* of places used for expressing of the conditions of a modeled process (we usually use circles for their representation), a finite set *T* of transitions describing the changes in the modeled process (we usually draw them in the form of rectangles) and a finite set *A* of arcs being principally oriented while connecting the place with the transition or the transition with the place and we usually draw them as lines with arrows.

Some commonly used notations for the nets are $\bullet y = \{x \mid (x, y) \in A\}$ for the **preset** and $y\bullet = \{x \mid (y, x) \in A\}$ for the **postset** of a net node *y* (i.e., place or transition). A **path** of a net *NET* := (*P*, *T*, *A*) is a nonempty sequence $<x_1, \ldots, x_k>$ of net nodes, where $k \in \mathbf{N}$, which satisfies $(x_1, x_2), (x_2, x_3), \ldots, (x_{k-1}, x_k) \in A$. A path of the net *NET* := (*P*, *T*, *A*) leading from its node *x* to its node *y* is a **circuit** if $(y, x) \in A$. We denote the set of all the circuits of the net *NET* by $\boldsymbol{CIRCUITS}_{NET}$. Net *NET′* := (*P′*, *T′*, *A′*) is a **subnet** of the net *NET* := (*P*, *T*, *A*) if $(P' \subseteq P) \wedge (T' \subseteq T) \wedge (A' = A \cap ((P' \times T') \cup (T' \times P')))$. Net *NET* is **connected** if and only if it is not composed of two or more disjoint and nonempty subnets.

Definition 3. Process net with time stamps (PNTS) *PNTS* is an ordered tuple *PNTS* := (*P*, *T*, *A*, *AF*, *TP*, *TI*, *IP*, *OP*, *RP*), where

i. (*P*, *T*, *A*) is the **connected net**, $\forall t \in T: (\bullet t \neq \varnothing) \wedge (t\bullet \neq \varnothing)$,

ii. $AF: (P \times T) \cup (T \times P) \rightarrow \mathbf{N}_0$ is the **arc function**,

$AF(x, y) > 0 \Leftrightarrow (x, y) \in A, AF(x, y) = 0 \Leftrightarrow (x, y) \notin A, \text{ where } x, y \in P \cup T,$

iii. *TP* is the **transition priority function**, $TP: T \rightarrow \mathbf{N}$,

iv. $TI: (T \times P) \rightarrow \mathbf{N}_0$ is the **time interval function**,

v. *IP* is the **input place**, $(IP \in P) \wedge (\bullet IP = \emptyset) \wedge (\forall p \in (P \setminus (\{IP\} \cup RP)): \bullet p \neq \emptyset)$,

vi. *OP* is the **output place**, $(OP \in P) \wedge (OP\bullet = \emptyset) \wedge (\forall p \in (P \setminus (\{OP\} \cup RP)): p\bullet \neq \emptyset)$,

vii. *RP* is the set of **resource places**, $(RP \subseteq (P \setminus \{IP, OP\})) \wedge (\forall t \in T: \neg(\bullet t \subseteq RP))$.

The class of all the PNTS will be denoted by ***PNTS***. □

The given PNTS $PNTS := (P, T, A, AF, TP, TI, IP, OP, RP)$ is represented by the **connected net** (P, T, A) with nonempty preset and postset of each of its transitions *t*; the **arc function** *AF* assigning each arc with a natural number (such number has the default value of 1, if not explicitly indicated in the PNTS diagram) expressing the number of removed or added tokens from or to the place associated with that arc when firing a particular transition; **transition priority function** *TP* assigns with each transition the natural number value expressing its priority (with the default value of 1); the **time interval function** *TI* assigns to each arc of the type (*transition, place*) a non-negative integer *d* expressing the minimum time interval during which the token has to remain in the *place* instead of being able to participate in the next firing of some transition and it thus determines the so-called **time marking** of the given PNTS (the value *d* associated with the respective arc is given in the format +*d* in the PNTS diagram); the **input place** *IP* is the only one nonresource place of PNTS *PNTS* with no input arc(s); the **output place** *OP* is the only one nonresource place of PNTS *PNTS* with no output arc(s); the finite set *RP* of **resource places** is used for expressing conditions of a modeled process containing some initial resources and we use circles with the double line for their representation.

Definition 4. Let $PNTS := (P, T, A, AF, TP, TI, IP, OP, RP)$ be the PNTS. Then:

i. **marking** *M* of the PNTS *PNTS* is a function $M: P \rightarrow \mathbf{N}_0$,

ii. **time marking** *m* of the PNTS *PNTS* is a function $m: P \rightarrow \mathbf{N}^{\#}$,

where $\forall p \in P: |M(p)| = length(m(p))$,

iii. variable $\tau \in \mathbf{N}_0$ is the **net time** of the PNTS *PNTS*,

iv. **state** *S* of the PNTS *PNTS* is an ordered triple $S := (M, m, \tau)$,

v. transition $t \in T$ is **enabled** in the state $S := (M, m, \tau)$ of the PNTS *PNTS* that is denoted by *t* ***en*** *S*, if $\forall p \in \bullet t: (M(p) \geq AF(p, t)) \wedge (\forall n \in elements(prefix(m(p), AF(p, t))): n \leq \tau))$,

vi. **firing of the transition** $t \in T$ results in changing the state $S := (M, m, \tau)$ of the PNTS *PNTS* into its state $S' := (M', m', \tau)$ that is denoted by $S\ [t\rangle\ S'$, where $\forall p \in P$:

- $M'(p) := M(p) - AF(p, t) + AF(t, p)$,
- $m'(p) := sort(suffix(m(p), AF(p, t)){+}{+}create(\tau + TI(t, p), AF(t, p)))$,

vii. **elapsing of time interval** $\delta \in \mathbf{N}$ results in changing the state $S := (M, m, \tau)$ of PNTS *PNTS* into its state $S' := (M, m, \tau + \delta)$, where $\forall t \in T: \neg(t\ \mathbf{en}\ (M, m, \tau))$, that is denoted by $S\ [\delta\rangle\ S'$, so that:

$$(\forall t \in T \, \forall n \in \mathbf{N}, 1 \leq n < \delta : \neg(t \; \mathbf{en} \; (M, m, \tau + n))) \wedge (\exists t \in T : t \; \mathbf{en} \; (M, m, \tau + \delta)),$$

viii. if the transitions $t_1, t_2, \ldots, t_n \in T$ are enabled in the state $S := (M, m, \tau)$ of PNTS *PNTS* (i.e., $(t_1 \; \mathbf{en} \; S) \wedge (t_2 \; \mathbf{en} \; S) \wedge \ldots \wedge (t_n \; \mathbf{en} \; S)$) we say that these transitions are **enabled in parallel** in the state S that is denoted by $\{t_1, t_2, \ldots, t_n\} \; \mathbf{en} \; S$,

ix. finite nonempty sequence $\sigma := t_1 \, t_2 \ldots t_n$ of the transitions $t_1, t_2, \ldots, t_n \in T$ for which the following is valid in the state $S_1 := (M_1, m_1, \tau_1)$ of PNTS *PNTS*:

- $(M_1, m_1, \tau_1) \, [t_1\rangle \, (M_2, m_2, \tau_1) \, [t_2\rangle \ldots [t_n\rangle \, (M_{n+1}, m_{n+1}, \tau_1)$,
- $\forall t \in T: \neg(t \; \mathbf{en} \; (M_{n+1}, m_{n+1}, \tau_1))$,

is called **step** σ in the given state S_1 of PNTS *PNTS* and it is denoted by.

$$(M_1, m_1, \tau_1) \, [\sigma\rangle \, (M_{n+1}, m_{n+1}, \tau_1),$$

x. finite nonempty sequence ρ of steps and time intervals elapsing that represents the following finite sequence

$$(M_1, m_1, \tau_1) \, [\sigma_1\rangle \, (M_2, m_2, \tau_1) \, [\delta_1\rangle \ldots (M_{n+1}, m_{n+1}, \tau_n) \, [\delta_n\rangle \, (M_{n+1}, m_{n+1}, \tau_{n+1})$$

of the state changes of PNTS *PNTS* is the sequence $\rho := \sigma_1 \, \delta_1 \, \sigma_2 \, \delta_2 \ldots \sigma_n \, \delta_n$ of steps $\sigma_1, \sigma_2, \ldots, \sigma_n$ and time intervals elapsing $\delta_1, \delta_2, \ldots, \delta_n$,

xi. we say the state S' of PNTS *PNTS* is reachable from its state S if there exists the finite sequence $\rho := \sigma_1 \, \delta_1 \, \sigma_2 \, \delta_2 \ldots \sigma_n \, \delta_n$ of steps $\sigma_1, \sigma_2, \ldots, \sigma_n$ and time intervals elapsing $\delta_1, \delta_2, \ldots, \delta_n$ such that $S \, [\sigma_1 \, \delta_1 \, \sigma_2 \, \delta_2 \ldots \sigma_n \, \delta_n\rangle \, S'$; the set of all the reachable states of PNTS *PNTS* from its state S is denoted by $[S\rangle$; the set of all the finite sequences $\rho := \sigma_1 \, \delta_1 \, \sigma_2 \, \delta_2 \ldots \sigma_n \, \delta_n$ associated with all the reachable states $S' \in [S\rangle$ is denoted by $[S\rangle\rangle$, that is,

$$[S\rangle\rangle := \{\sigma_1 \, \delta_1 \, \sigma_2 \, \delta_2 \ldots \sigma_n \, \delta_n \mid \exists S' \in [S\rangle : S \, [\sigma_1 \, \delta_1 \, \sigma_2 \, \delta_2 \ldots \sigma_n \, \delta_n\rangle \, S', n \in \mathbf{N}\},$$

xii. the set of all the states $S := (M, m, \tau)$ of PNTS *PNTS* is denoted by $\mathbf{S}$,

xiii. the set of all the markings M associated with the set $\mathbf{S}$ of all the states of PNTS *PNTS* is denoted by $\boldsymbol{M}$, that is, $\boldsymbol{M} := \{M \mid (S = (M, m, \tau)) \wedge (S \in \mathbf{S})\}$,

xiv. **static state** $S_s := (M_s, m_s, \tau_s)$ of PNTS *PNTS* is every of its states where

$$\forall p \in P \backslash RP : (M_s(p) = 0) \wedge (m_s(p) = \langle\rangle),$$

xv. the set of all the static states $S_s := (M_s, m_s, \tau_s)$ of PNTS *PNTS* is denoted by $\mathbf{S_s}$,

xvi. the set of all the static markings M_s associated with the set $\mathbf{S_s}$ of all the static states of PNTS *PNTS* is denoted by $\boldsymbol{M_s}$, that is, $\boldsymbol{M_s} := \{M_s \mid (S_s := (M_s, m_s, \tau_s)) \wedge (S_s \in \mathbf{S_s})\}$,

xvii. the function $\xi: \boldsymbol{M} \rightarrow \boldsymbol{M_s}$ which assigns to each marking $M \in \boldsymbol{M}$ of a given PNTS *PNTS* the associated static marking $M_s \in \boldsymbol{M_s}$ is defined as follows:

- $\forall p \in RP: \xi(M(p)) := M(p)$,
- $\forall p \in P \setminus RP: \xi(M(p)) := 0$,

xviii. **entry state** $S_e := (M_e, m_e, \tau_e)$ of PNTS *PNTS* is every of its states where

- $k \in N: (Me(IP) = k) \wedge (length(me(IP)) = k)$,
- $\forall p \in P \setminus (RP \cup \{IP\}): (Me(p) = 0) \wedge (me(p) = <>)$,
- $\forall p \in RP: (Me(p) \geq 0) \wedge (me(p) \in N^0)$,

xix. the set of all the entry states $S_e := (M_e, m_e, \tau_e)$ of PNTS *PNTS* is denoted by $\mathbf{S_e}$,

xx. **exit state** $S_x := (M_x, m_x, \tau_x)$ of PNTS *PNTS* that is reachable from its entry state $S_e := (M_e, m_e, \tau_e)$ is every of its states where

- $S_x \in [S_e\rangle$,
- $M_x(OP) = M_e(IP)$,
- $\forall p \in P \setminus (RP \cup \{OP\}): (M_e(p) = 0) \wedge (m_e(p) = <>)$,

xxi. the set of all the exit states $S_x := (M_e, m_e, \tau_e)$ of PNTS *PNTS* that are reachable from its entry state $S_e := (M_e, m_e, \tau_e)$ is denoted by $[S_e\rangle_x$,

xxii. the set of all the exit states S_x of PNTS *PNTS* that are reachable from all its entry states $S_e \in \mathbf{S_e}$ is denoted by $\mathbf{S_x}$. □

The above established concepts are demonstrated in a simple example of the PNTS *PNTS*1 := (*P*, *T*, *A*, *AF*, *TP*, *TI*, *IP*, *OP*, *RP*) that is shown in **Figure 1**, where *P* := {**IP**, **P1**, **R1**, **OP**}, *T* := {**T1**, **T2**, **T3**}, *A* := {(**IP**, **T1**), (**IP**, **T2**), (**T1**, **P1**), (**T2**, **P1**), (**R1**, **T1**), (**P1**, **T3**), (**T3**, **R1**), (**T3**, **OP**)}, *AF* := {((**IP**, **T1**), 1), ((**IP**, **T2**), 1), ((**T1**, **P1**), 1), ((**T2**, **P1**), 2), ((**R1**, **T1**), 1), ((**P1**, **T3**), 1), ((**T3**, **R1**), 1), ((**T3**, **OP**), 1)}, *TP* := {(**T1**, 2), (**T2**, 1), (**T3**, 1)}, *TI* := {((**T1**, **P1**), 3), ((**T2**, **P1**), 3), ((**T3**, **R1**), 1), ((**T3**, **OP**), 4)}, *IP* := **IP**, *OP* := **OP**, *RP* := {**R1**}.

PNTS *PNTS*1 is in its entry state $S_e := (M_e, m_e, \tau_e)$, where marking $M_e := (M_e(\mathbf{IP}), M_e(\mathbf{P_1}), M_e(\mathbf{R1}), M_e(\mathbf{OP})) = (2, 0, 2, 0)$, time marking $m_e := (m_e(\mathbf{IP}), m_e(\mathbf{P_1}), m_e(\mathbf{R1}), m_e(\mathbf{OP})) = (<0, 2>, <>, <0, 0>, <>)$ and net time $\tau_e = 0$ (i.e., $\tau_e = \tau$). Static marking $M_s \in \mathbf{M_s}$ associated with the entry marking M_e (see (xiv) and (xvii) of Definition 4) has the value $M_s := \xi(M_e) = (0, 0, 2, 0)$.

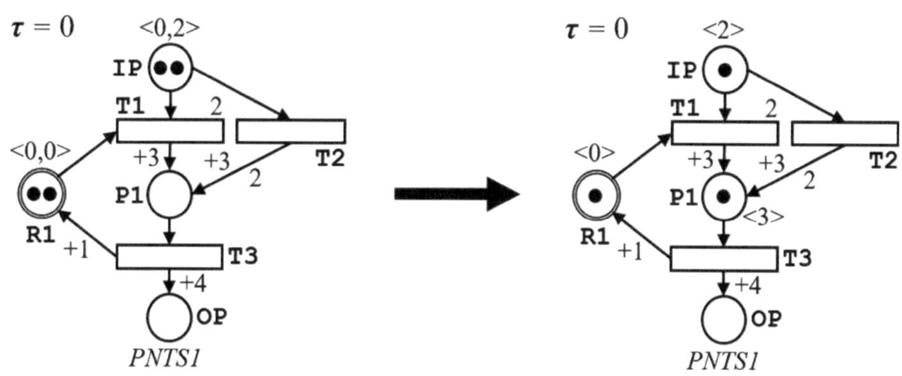

Figure 1. Firing of transition **T1** in PNTS *PNTS*1.

Time marking m of any PNTS expresses the current time state of the modeled system using the final (or empty) ascending ordered sequences of non-negative integers (i.e., elements of the set $\mathbf{N}^{\#}$) associated with each of its places. The individual values of the time marking m associated with the arbitrary place p of the given PNTS in its state S, informally said, represent the values of the net time τ at which the respective token can first participate in the firing of selected enabled transition t of the given PNTS.

The transitions **T1** and **T2** are enabled in the entry state S_e because (see (v) of Definition 4):

- $\forall p \in \bullet\mathbf{T1}$: $(2 = M(\mathbf{IP}) \geq AF(\mathbf{IP}, \mathbf{T1}) = 1) \wedge (2 = M(\mathbf{R1}) \geq AF(\mathbf{R1}, \mathbf{T1}) = 1) \wedge (\forall n \in elements(prefix(m(\mathbf{IP}), AF(\mathbf{IP}, \mathbf{T1}))) = elements(prefix(<0, 2>, 1)) = elements(<0>) = \{0\}: 0 \leq 0) \wedge (\forall n \in elements(prefix(m(\mathbf{R1}), AF(\mathbf{R1}, \mathbf{T1}))) = elements(prefix(<0, 0>, 1)) = elements(<0>) = \{0\}: 0 \leq 0)$,
- $\forall p \in \bullet\mathbf{T2}$: $(2 = M(\mathbf{IP}) \geq AF(\mathbf{IP}, \mathbf{T1}) = 1) \wedge (\forall n \in elements(prefix(m(\mathbf{IP}), AF(\mathbf{IP}, \mathbf{T2}))) = elements(prefix(<0, 2>, 1)) = elements(<0>) = \{0\}: 0 \leq 0)$.

When enabling individual transitions of the given PNTS so-called **conflicts** can originate in its certain markings (or **conflict transitions**). At the enabling of the transitions t_1 and t_2 of the given PNTS in its state S the conflict occurs, if both transitions t_1 and t_2 have at least one input place, each of the transitions t_1 and t_2 is individually enabled in the state S, but the transitions t_1 and t_2 are not enabled in parallel in the state S (see (viii) of Definition 4) and enabling of one of them will prevent enabling of the other (i.e., $(\bullet t_1 \cap \bullet t_2 \neq \emptyset) \wedge (t_1\ \mathbf{en}\ S) \wedge (t_2\ \mathbf{en}\ S) \wedge \neg(\{t_1, t_2\}\ \mathbf{en}\ S)$). The term of conflict transitions can be obviously easily generalized for the case of a finite set $t_1, t_2, \ldots, t_n, n \in \mathbf{N}$ of the transitions of the given PNTS.

The transitions **T1** and **T2** in the entry state S_e of PNTS *PNTS*1 are conflict transitions because the time marking $m_e(\mathbf{IP}) = <0, 2>$ (i.e., only one token of the entry marking $M_e(\mathbf{IP})$ may participate in the firing of the transition **T1** or **T2** in the net time $\tau_e = 0$). When solving such transitions conflict we therefore follow the rule which determines, informally said, that from the set of conflict transitions the one will be enabled, whose value of the transition priority function *TP* is the highest. If such transition from the set of conflict transitions does not exist, the given conflict would have to be solved by other means. The transition **T1** is then enabled in the entry state S_e on the basis of that rule in our studied example (because $TP(\mathbf{T1}) = 2$ and $TP(\mathbf{T2}) = 1$).

Firing of the transition **T1** changes the entry state $S_e := (M_e, m_e, \tau_e)$ of the PNTS *PNTS*1 into its state $S_1 := (M_1, m_1, \tau_e)$ (i.e., $S_e\ [\mathbf{T1}\rangle\ S_1$—see **Figure 1**), where (see (vi) of Definition 4):

- $M_1(\mathbf{IP}) := M_e(\mathbf{IP}) - AF(\mathbf{IP}, \mathbf{T1}) = 2 - 1 = 1$,
- $m_1(\mathbf{IP}) := sort(suffix(m_e(\mathbf{IP}), AF(\mathbf{IP}, \mathbf{T1}))) = sort(suffix(<0, 2>, 1)) = sort(<2>) = <2>$,
- $M_1(\mathbf{P1}) := M_e(\mathbf{P1}) + AF(\mathbf{T1}, \mathbf{P1}) = 0 + 1 = 1$,
- $m_1(\mathbf{P1}) := sort(create(\tau + TI(\mathbf{T1}, \mathbf{P1}), AF(\mathbf{T1}, \mathbf{P1}))) = sort(create(0 + 3, 1)) = sort(create(3, 1)) = sort(<3>) = <3>$,
- $M_1(\mathbf{R1}) := M_e(\mathbf{R1}) - AF(\mathbf{R1}, \mathbf{T1}) = 2 - 1 = 1$,
- $m_1(\mathbf{R1}) := sort(suffix(m_e(\mathbf{R1}), AF(\mathbf{R1}, \mathbf{T1}))) = sort(suffix(<0, 0>, 1)) = sort(<0>) = <0>$.

There is no enabled transition in the state $S_1 := (M_1, m_1, \tau_1)$ and it is necessary to perform the time interval elapsing with the value of $\delta = 2$. This will change the state $S_1 := (M_1, m_1, \tau_e)$ into the state $S_2 := (M_1, m_1, \tau_1)$, where $\tau_1 := \tau_e + \delta = 2$ (i.e., S_1 [2⟩ S_2). It can be easily shown that transition **T1** in the state S_2 is enabled and firing of this transition changes the state $S_2 := (M_1, m_1, \tau_1)$ of the PNTS *PNTS*1 into its state $S_3 := (M_2, m_2, \tau_1)$ (i.e., S_2 [**T1**⟩ S_3), where $M_2 := (M_2(\mathbf{IP}), M_2(\mathbf{P1}), M_2(\mathbf{R1}), M_2(\mathbf{OP})) = (0, 2, 0, 0)$ and $m_2 := (m_2(\mathbf{IP}), m_2(\mathbf{P1}), m_2(\mathbf{R1}), m_2(\mathbf{OP})) = (\langle\rangle, \langle 3, 5\rangle, \langle\rangle, \langle\rangle)$.

It can then be easily verified that S_3 [1⟩ S_4 [**T3**⟩ S_5 [2⟩ S_6 [**T3**⟩ S_x, where:

- $S_4 = (M_2, m_2, \tau_2) = ((0, 2, 0, 0), (\langle\rangle, \langle 3, 5\rangle, \langle\rangle, \langle\rangle), 3)$,
- $S_5 = (M_3, m_3, \tau_2) = ((0, 1, 1, 1), (\langle\rangle, \langle 5\rangle, \langle 4\rangle, \langle 7\rangle), 3)$,
- $S_6 = (M_3, m_3, \tau_3) = ((0, 1, 1, 1), (\langle\rangle, \langle 5\rangle, \langle 4\rangle, \langle 7\rangle), 5)$,
- $S_x = (M_4, m_4, \tau_3) = ((0, 0, 2, 2), (\langle\rangle, \langle\rangle, \langle 4, 6\rangle, \langle 7, 9\rangle), 5)$.

There are no enabled transitions in the exit state $S_x := (M_4, m_4, \tau_3)$ of PNTS *PNTS*1 that is reachable from the entry state $S_e := (M_e, m_e, \tau_e)$ (see (xx) of Definition 4) and there is also no time interval elapsing value δ in this state that enables any of the transitions.

The set $\boldsymbol{AM_s} \subseteq \boldsymbol{M_s}$ of all the **allowed static markings** of the given PNTS *PNTS*, informally said, expresses how many tokens may be located in its individual resource places if PNTS *PNTS* be in its (now no longer arbitrary) allowed entry state $AS_e \in \boldsymbol{AS_e}$, where $\boldsymbol{AS_e} \subseteq \boldsymbol{S_e}$. For instance, the set $\boldsymbol{AM_s}$ of the PNTS *PNTS*1 (see **Figure 1**) can be defined as $\boldsymbol{AM_s} := \{(0, 0, k, 0) \mid k \in \mathbf{N}\}$, that is, there must be at least one token in the resource place **R1** (and of course at least one token in the input place **IP**) in any allowed entry state $AS_e \in \boldsymbol{AS_e}$.

Definition 5. Let $PNTS := (P, T, A, AF, TP, TI, IP, OP, RP)$ be a PNTS, $\boldsymbol{AM_s} \subseteq \boldsymbol{M_s}$ be the set of all of its allowed static markings and $\boldsymbol{AS_e} := \{AS_e \mid (AS_e = (AM_e, am_e, 0)) \wedge (\xi(AM_e) \in \boldsymbol{AM_s})\}$ be the set of all of its allowed entry states. Then:

i. *PNTS* is *k*-**bounded** PNTS if

$$\forall AS_e \in \boldsymbol{AS_e}\, \exists k \in \mathbf{N}_0 \forall p \in P\, \forall S \in [AS_e\rangle, S{:=}(M, m, \tau) : M(p) \le k,$$

ii. *PNTS* is **proper-formed** PNTS if

$$\forall AS_e \in \boldsymbol{AS_e} : (\forall S \in [AS_e\rangle \exists S_x \in [AS_e\rangle_x : S_x \in [S\rangle) \wedge (\exists n \in \mathbf{N} : |[AS_e\rangle\rangle| = n),$$

iii. proper-formed *PNTS* is **well-formed** PNTS if

$$\forall AS_e \in \boldsymbol{AS_e} \forall S_x \in [AS_e\rangle_x, S_x{:=}(M_x, m_x, \tau_x) : \xi(M_x) \in \boldsymbol{AM_s},$$

iv. well-formed *PNTS* is **pure-formed** PNTS if

$$\forall AS_e \in \boldsymbol{AS_e} \forall S_x \in [AS_e\rangle_x, S_x{:=}(M_x, m_x, \tau_x) : \xi(AM_e) = \xi(M_x).$$ □

PNTS is **proper-formed** PNTS if for any of its state S that is reachable from any allowed entry state $AS_e \in \boldsymbol{AS_e}$ there exists its output state S_x that is also reachable from its allowed entry state

$AS_e \in \boldsymbol{AS_e}$ such that the output state S_x is also reachable from the state S (i.e., $\forall S \in [AS_e\rangle\ \exists S_x \in [AS_e\rangle_x$: $S_x \in [S\rangle$). Furthermore, the cardinality of the set $[AS_e\rangle\rangle$ of all the sequences $\rho := \sigma_1\ \delta_1\ \sigma_2\ \delta_2\ ...\ \sigma_n\ \delta_n$ associated with all the reachable states $S \in [AS_e\rangle\rangle$ must be finite (i.e., ($\exists n \in \mathbf{N}$: $|[AS_e\rangle\rangle| = n$).

Proper-formed *PNTS* is **well-formed** PNTS if for any of its allowed entry state $AS_e \in \boldsymbol{AS_e}$ and for any of its exit state $S_x \in [AS_e\rangle_x$, where $S_x := (M_x, m_x, \tau_x)$, it is true that the exit static marking $\xi(M_x)$ of all its resource places **is an element** of the set $\boldsymbol{AM_s}$ of all its allowed static markings if PNTS *PNTS* be in its allowed entry state $AS_e \in \boldsymbol{AS_e}$ (i.e., $\forall AS_e \in \boldsymbol{AS_e}\ \forall S_x \in [AS_e\rangle_x$, $S_x := (M_x, m_x, \tau_x)$: $\xi(M_x) \in \boldsymbol{AM_s}$).

Well-formed *PNTS* is **pure-formed** PNTS if for any of its allowed entry state $AS_e \in \boldsymbol{AS_e}$, where $AS_e := (AM_e, am_e, \tau_e)$, and for any of its exit state $S_x \in [AS_e\rangle_x$, where $S_x := (M_x, m_x, \tau_x)$, it is true that the exit static marking $\xi(M_x)$ of all its resource places **is equal to** the entry static marking $\xi(AM_e)$ of all its resource places that is associated with the allowed entry state AS_e (i.e., $\forall AS_e \in \boldsymbol{AS_e}\ \forall S_x \in [AS_e\rangle_x$, $S_x := (M_x, m_x, \tau_x)$: $\xi(AM_e) = \xi(M_x)$).

For instance, if the set $\boldsymbol{AM_s}$ of the PNTS *PNTS*1 (see **Figure 1**) is defined as:

i. $\boldsymbol{AM_s} := \{(0, 0, k, 0) \mid k \in \mathbf{N}\}$ (i.e., there must be at least one token in the resource place **R1** in any allowed entry state $AS_e \in \boldsymbol{AS_e}$), then it can be shown that PNTS *PNTS*1 is *k*-bounded, proper-formed, well-formed and pure-formed PNTS,

ii. $\boldsymbol{AM_s} := \{(0, 0, 0, 0)\}$ (i.e., there may not be any token in the resource place **R1** in any allowed entry state $AS_e \in \boldsymbol{AS_e}$), then it can be shown that PNTS *PNTS*1 is *k*-bounded, proper-formed, but not well-formed or pure-formed PNTS (see for instance the sequence

$((1,0,0,0),(<0>,<>,<>,<>),0)\ [\mathbf{T2}\rangle\ ((0,2,0,0),(<>,<3,3>,<>,<>),0)\ [3\rangle.$
$((0,2,0,0),(<>,<3,3>,<>,<>),3)\ [\mathbf{T3\ T3}\rangle\ ((0,0,2,2),(<>,<>,<4,4>,<7,7>),3),$
$where\ \xi(M_x) = \xi((0,0,2,2)) = (0,0,2,0) \notin \{(0,0,0,0)\} = \boldsymbol{AM_s}).$

Lemma 1. If *PNTS* is proper-formed PTNS then *PNTS* is *k*-bounded PNTS.

Proof. Clear. PNTS *PNTS* := (*P*, *T*, *A*, *AF*, *TP*, *TI*, *IP*, *OP*, *RP*) is a connected net that contains the finite set *T* of the transitions. Then the finite number of tokens will be added to each of the places $p \in P$ by firing each of the transitions $t \in T$. The number of states $S \in [AS_e\rangle$ for any allowed entry state $AS_e \in \boldsymbol{AS_e}$ must be also finite because *PNTS* is proper-formed PNTS (i.e., $\exists n \in \mathbf{N}$: $|[AS_e\rangle\rangle| = n$). From these facts then immediately follows that in any state $S \in [AS_e\rangle$ the finite number of tokens must be placed in any place $p \in P$, where any final number of tokens is placed in the input place *IP* in the entry state AS_e. From these facts then immediately follows that $\forall AS_e \in \boldsymbol{AS_e}\ \exists k \in \mathbf{N}_0\ \forall p \in P\ \forall S \in [AS_e\rangle, S := (M, m, \tau) : M(p) \leq k$. □

Definition 6. Process Petri net with time stamps (PPNTS) *PPNTS* is an ordered couple *PPNTS* := (*PNTS*, S_e), where *PNTS* := (*P*, *T*, *A*, *AF*, *TP*, *TI*, *IP*, *OP*, *RP*) is the PNTS and $S_e \in \boldsymbol{S_e}$ is the entry state of PNTS *PNTS*. The class of all PPNTSs will be denoted by ***PPNTS***. □

3. Algebraic operators COMP and SYNC and their properties

We study the issue of transforming PNTS through precisely defined binary operator **COMP** and n-ary operator **SYNC** over the class ***PNTS*** and we also examine the preservation of individual PNTS's properties when applying each of these operators. Formal enrollment of an application of generally n-ary operator **OP** whose operands are the PNTS $PNTS_1$, $PNTS_2$, …, $PNTS_n$ ($n \in \mathbf{N}$) and whose application requires the specification of values of k formal parameters ($k \in \mathbf{N}$) par_1, par_2, … par_k, will be denoted by the expression.

$$PNTS{:=}[PNTS_1, PNTS_2, \dots, PNTS_n].\mathbf{OP}(par_1, par_2, \dots, par_k),$$

where $PNTS$ is the resulting PNTS.

Definition 7. Let $PNTS_1 := (P_1, T_1, A_1, AF_1, TP_1, TI_1, IP_1, OP_1, RP_1)$ and $PNTS_2 := (P_2, T_2, A_2, AF_2, TP_2, TI_2, IP_2, OP_2, RP_2)$ be the PNTSs. Let $\boldsymbol{AM}_{s1} := \{(AM_{s1}(IP_1), AM_{s1}(p1_1), \dots, AM_{s1}(p1_n), AM_{s1}(r1_1), \dots, AM_{s1}(r1_m), AM_{s1}(OP_1)) \mid P_1 := \{p1_1, \dots, p1_n, r1_1, \dots, r1_m\}, RP_1 := \{r1_1, \dots, r1_m\}, n \in \mathbf{N}, m \in \mathbf{N}\}$ be the set of all the allowed static markings of $PNTS_1$, $\boldsymbol{AM}_{s2} := \{(AM_{s2}(IP_2), AM_{s2}(p2_1), \dots, AM_{s2}(p2_k), AM_{s2}(r2_1), \dots, AM_{s2}(r2_h), AM_{s2}(OP_2)) \mid P_2 := \{p2_1, \dots, p2_k, r2_1, \dots, r2_h\}, RP_2 := \{r2_1, \dots, r2_h\}, k \in \mathbf{N}, h \in \mathbf{N}\}$ be the set of all the allowed static markings of $PNTS_2$.

Cartesian product $\boldsymbol{AM}_{s1} \otimes \boldsymbol{AM}_{s2}$ is then the following set:

$$\begin{gathered}\boldsymbol{AM}_{s1} \otimes \boldsymbol{AM}_{s2}{:=}\{(AM_{s1}(IP_1), AM_{s1}(p1_1), \dots, AM_{s1}(p1_n), AM_{s1}(r1_1), \dots, AM_{s1}(r1_m), AM_{s1}(OP_1),\\ AM_{s2}(IP_2), AM_{s2}(p2_1), \dots, AM_{s2}(p2_k), AM_{s2}(r2_1), \dots, AM_{s2}(r2_h), AM_{s2}(OP_2)) \mid.\\ (AM_{s1}(IP_1), AM_{s1}(p1_1), \dots, AM_{s1}(p1_n), AM_{s1}(r1_1), \dots, AM_{s1}(r1_m), AM_{s1}(OP_1)) \in \boldsymbol{AM}_{s1} \wedge.\\ (AM_{s2}(IP_2), AM_{s2}(p2_1), \dots, AM_{s2}(p2_k), AM_{s2}(r2_1), \dots, AM_{s2}(r2_h), AM_{s2}(OP_2)) \in \boldsymbol{AM}_{s2}\}.\end{gathered}$$

PNTS $PNTS_1$ and $PNTS_2$ are disjoint and we denote this fact by $PNTS_1 \angle PNTS_2$ if.

$$(P_1 \cap P_2 = \varnothing) \wedge (T_1 \cap T_2 = \varnothing). \qquad \square$$

Definition 8. The function **COMP**: ***PNTS*** $\times$ ***PNTS*** $\to$ ***PNTS* of nets composition** is defined as follows: if $PNTS_1 := (P_1, T_1, A_1, AF_1, TP_1, TI_1, IP_1, OP_1, RP_1)$ and $PNTS_2 := (P_2, T_2, A_2, AF_2, TP_2, TI_2, IP_2, OP_2, RP_2)$ be the arbitrary PNTSs, $PNTS_1 \angle PNTS_2$, t be an arbitrary transition, where $(t \notin T_1) \wedge (t \notin T_2)$, $ti \in \mathbf{N}_0$, then $PNTS := [PNTS_1, PNTS_2]$.**COMP**(t, ti), where PNTS $PNTS := (P, T, A, AF, TP, TI, IP, OP, RP)$ fulfills the following:

i. $P := P_1 \cup P_2$,

ii. $T := T_1 \cup T_2 \cup \{t\}$,

iii. $A := A_1 \cup A_2 \cup \{(OP_1, t), (t, IP_2)\}$,

iv. $AF := AF_1 \cup AF_2 \cup \{((OP_1, t), 1), ((t, IP_2), 1)\}$,

v. $TP := TP_1 \cup TP_2 \cup \{(t, 1)\}$,

vi. $TI := TI_1 \cup TI_2 \cup \{(t, IP_2), ti)\}$,

vii. $IP := IP_1$,

viii. $OP := OP_2$,

ix. $RP := RP_1 \cup RP_2$. □

Symbolic representation of PNTS $[PNTS_1, PNTS_2]$.**COMP**(t, ti) can be seen in **Figure 2**.

Lemma 2. Let $PNTS_1 := (P_1, T_1, A_1, AF_1, TP_1, TI_1, IP_1, OP_1, RP_1)$ and $PNTS_2 := (P_2, T_2, A_2, AF_2, TP_2, TI_2, IP_2, OP_2, RP_2)$ be two arbitrary PNTS, $PNTS_1 \angle PNTS_2$, t be an arbitrary transition, $(t \notin T_1) \wedge (t \notin T_2)$, $ti \in \mathbf{N}_0$, $\boldsymbol{AM}_{s1}$ and $\boldsymbol{AM}_{s2}$ be the sets of all the allowed static markings of $PNTS_1$ and $PNTS_2$. Let $PNTS := [PNTS_1, PNTS_2]$.**COMP**(t, ti).

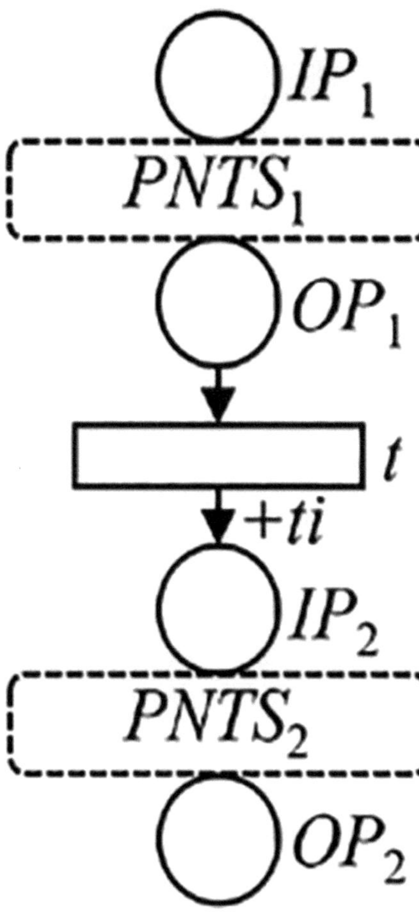

Figure 2. Symbolic representation of PNTS $[PNTS_1, PNTS_2]$.**COMP**(t, ti).

If $PNTS_1$ and $PNTS_2$ are proper-formed, resp. well-formed, resp. pure-formed, PNTS and $\boldsymbol{AM_s} = \boldsymbol{AM_{s1}} \otimes \boldsymbol{AM_{s2}}$ be the set of all the allowed static markings of PNTS *PNTS*, then also resulting *PNTS* is proper-formed, resp. well-formed, resp. pure-formed, PNTS.

Proof. Clear, it directly follows from Definition 5, Definition 7 and Definition 8. □

Definition 9. The function **SYNC**: $\boldsymbol{PNTS} \times \boldsymbol{PNTS} \times \ldots \times \boldsymbol{PNTS} \rightarrow \boldsymbol{PNTS}$ **of synchronous nets composition** is defined as follows: if $PNTS_1 := (P_1, T_1, A_1, AF_1, TP_1, TI_1, IP_1, OP_1, RP_1)$, $PNTS_2 := (P_2, T_2, A_2, AF_2, TP_2, TI_2, IP_2, OP_2, RP_2)$, …, $PNTS_n := (P_n, T_n, A_n, AF_n, TP_n, TI_n, IP_n, OP_n, RP_n)$, be the arbitrary PNTSs, $\forall i, 1 \le i \le n, \forall j, 1 \le j \le n: i \neq j \Rightarrow PNTS_i \angle PNTS_j$, where $n \in \mathbf{N}$, *pi* and *po* be the arbitrary places, $(pi \notin P_1 \cup P_2 \cup \ldots \cup P_n) \wedge (po \notin P_1 \cup P_2 \cup \ldots \cup P_n) \wedge (pi \neq po)$, *ti* and *to* be the arbitrary transitions, $(ti \notin T_1 \cup T_2 \cup \ldots \cup T_n) \wedge (to \notin T_1 \cup T_2 \cup \ldots \cup T_n) \wedge (ti \neq to)$, $af1 \in \mathbf{N}$, $af2 \in \mathbf{N}$, …, $afn \in \mathbf{N}$, $ti1 \in \mathbf{N}_0$, $ti2 \in \mathbf{N}_0$, …, $tin \in \mathbf{N}_0$, $tio \in \mathbf{N}_0$, then

$$PNTS := [PNTS_1, PNTS_2, \ldots, PNTS_n].\mathbf{SYNC}(pi, po, ti, to, af1, \ldots, afn, ti1, \ldots, tin, tio),$$

where PNTS $PNTS := (P, T, A, AF, TP, TI, IP, OP, RP)$ fulfills the following:

i. $P := P_1 \cup P_2 \cup \ldots \cup P_n \cup \{pi, po\}$,

ii. $T := T_1 \cup T_2 \cup \ldots \cup T_n \cup \{ti, to\}$,

iii. $A := A_1 \cup A_2 \cup \ldots \cup A_n \cup \{(pi, ti), (ti, IP_1), \ldots, (ti, IP_n), (OP_1, to), \ldots, (OP_n, to), (to, po)\}$,

iv. $AF := AF_1 \cup AF_2 \cup \ldots \cup AF_n \cup \{((pi, ti), 1), ((ti, IP_1), af1), \ldots, ((ti, IP_n), afn), ((OP_1, to), af1), \ldots, ((OP_n, to), afn), ((to, po), 1)\}$,

v. $TP := TP_1 \cup TP_2 \cup \ldots \cup TP_n \cup \{(ti, 1), (to, 1)\}$,

vi. $TI := TI_1 \cup TI_2 \cup \ldots \cup TI_n \cup \{((ti, IP_1), ti1), \ldots, ((ti, IP_n), tin), ((to, po), tio)\}$,

vii. $IP := pi$,

viii. $OP := po$,

ix. $RP := RP_1 \cup RP_2 \cup \ldots \cup RP_n$. □

Symbolic representation of PNTS $[PNTS_1, PNTS_2, \ldots, PNTS_n].\mathbf{SYNC}(pi, po, ti, to, af1, \ldots, afn, ti1, \ldots, tin, tio)$ can be seen in **Figure 3**.

Lemma 3. Let $PNTS_1 := (P_1, T_1, A_1, AF_1, TP_1, TI_1, IP_1, OP_1, RP_1)$, $PNTS_2 := (P_2, T_2, A_2, AF_2, TP_2, TI_2, IP_2, OP_2, RP_2)$, …, $PNTS_n := (P_n, T_n, A_n, AF_n, TP_n, TI_n, IP_n, OP_n, RP_n)$ be arbitrary PNTSs, $\forall i, 1 \le i \le n, \forall j, 1 \le j \le n: i \neq j \Rightarrow PNTS_i \angle PNTS_j$, where $n \in \mathbf{N}$, *pi* and *po* be arbitrary places, $(pi \notin P_1 \cup P_2 \cup \ldots \cup P_n) \wedge (po \notin P_1 \cup P_2 \cup \ldots \cup P_n) \wedge (pi \neq po)$, *ti* and *to* be arbitrary transitions, $(ti \notin T_1 \cup T_2 \cup \ldots \cup T_n) \wedge (to \notin T_1 \cup T_2 \cup \ldots \cup T_n) \wedge (ti \neq to)$, $af1 \in \mathbf{N}$, $af2 \in \mathbf{N}$, …, $afn \in \mathbf{N}$, $ti1 \in \mathbf{N}_0$, $ti2 \in \mathbf{N}_0$, …, $tin \in \mathbf{N}_0$, $tio \in \mathbf{N}_0$ and $\boldsymbol{AM_{s1}}, \boldsymbol{AM_{s2}}, \ldots, \boldsymbol{AM_{sn}}$ be the sets of all the allowed static markings of $PNTS_1, PNTS_2, \ldots, PNTS_n$. Let $PNTS := [PNTS_1, PNTS_2, \ldots, PNTS_n].\mathbf{SYNC}(pi, po, ti, to, af1, \ldots, afn, ti1, \ldots, tin, tio)$.

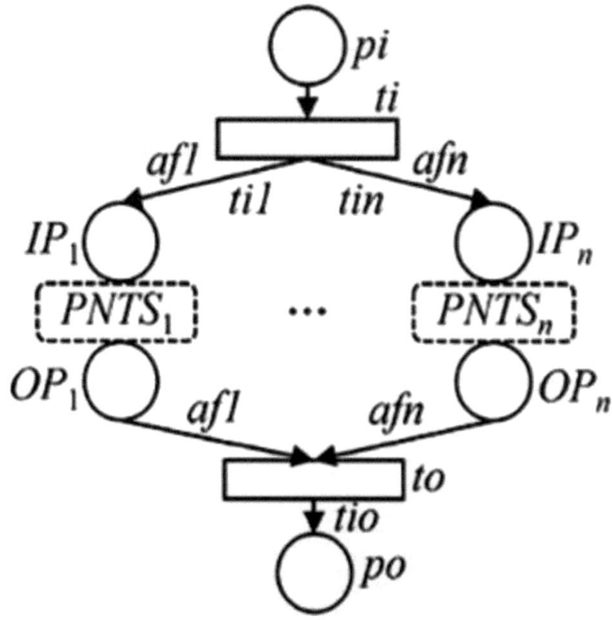

Figure 3. Symbolic representation of PNTS [$PNTS_1$, $PNTS_2$, …, $PNTS_n$].**SYNC**(*pi, po, ti, to, af1, …, afn, ti1, …, tin, tio*).

If $PNTS_1$, $PNTS_2$, …, $PNTS_n$ are proper-formed, resp. well-formed, resp. pure-formed, PNTS and $\boldsymbol{AM_s} = \boldsymbol{AM_{s1}} \otimes \boldsymbol{AM_{s2}} \otimes \ldots \otimes \boldsymbol{AM_{sn}}$ is the set of all the allowed static markings of PNTS *PNTS*, then also *PNTS* is proper-formed, resp. well-formed, resp. pure-formed, PNTS.

Proof. Clear, it directly follows from Definition 5, Definition 7 and Definition 9. □

4. Process Petri nets with time stamps and their applications in project management area

Critical Path Method (CPM) is a method used in modeling and project management that was developed at the end of 1950s and that is commonly used for all the types of projects including software development [10]. The CPM is the most widely used method of so-called network analysis, even though it is designed to analyze the time consumption of only deterministic projects, that is, projects where the duration of each of their activities is exactly known, including all their sub-activities.

The basis for using CPM is to create a project model that includes:

- the list of all activities needed to complete the project,
- the time duration of each activity that is constant,
- the dependencies between the project activities,

A critical path is then a designation for a sequence of activities whose time duration directly affects the time duration of the entire project. The activities that make up the critical path are then referred to as critical activities. There may be several critical paths in the project. When managing the project, a sequence of activities within a given network chart describing this project that increases the longest total time duration of a project is called its critical path. The critical path within the network chart can be used to determine the shortest time required to complete the project. The application of the CPM method can therefore determine which activities within the studied project are "critical" (i.e., activities on the longest path in the network chart describing the project) and which activities may be delayed in the execution of the project without increasing its total time.

The special class $\boldsymbol{CPNET} \subset \boldsymbol{PNTS}$ of PNTS is introduced in the following paragraphs to represent network chart used in the CPM method through PNTS. Special unary operator **JOIN** that is required in the definition of the class ***CPNET*** is introduced first.

Definition 10. The function **JOIN**: $\boldsymbol{PNTS} \rightarrow \boldsymbol{PNTS}$ **of net transition joining** is defined as follows: if $PNTS_1 := (P_1, T_1, A_1, AF_1, TP_1, TI_1, IP_1, OP_1, RP_1)$ be the arbitrary PNTS, $p \notin P_1$ be the arbitrary place, $t1$ and $t2$ be the arbitrary transitions, $(t1 \neq t2) \wedge (t1 \in T_1) \wedge (t2 \in T_1)$, $ti \in \mathbf{N}_0$, then $PNTS := PNTS_1.$**JOIN**$(p, t1, t2, ti)$, where PNTS $PNTS := (P, T, A, AF, TP, TI, IP, OP, RP)$ fulfills the following:

i. $P := P_1 \cup \{p\}$,

ii. $T := T_1$,

iii. $A := A_1 \cup \{(t1, p), (p, t2)\}$,

iv. $AF := AF_1 \cup \{((t1, p), 1), ((p, t2), 1)\}$,

v. $TP := TP_1 \cup \{(t, 1)\}$,

vi. $TI := TI_1 \cup \{((t1, p), ti)\}$,

vii. $IP := IP_1$,

viii. $OP := OP_1$,

ix. $RP := RP_1$. □

Symbolic representation of the unary operator **JOIN** application over the PNTS $PNTS_1$ can be seen in **Figure 4**.

Definition 11. Let $PNTS_1 := (P_1, T_1, A_1, AF_1, TP_1, TI_1, IP_1, OP_1, RP_1)$, $PNTS_2 := (P_2, T_2, A_2, AF_2, TP_2, TI_2, IP_2, OP_2, RP_2)$, …, $PNTS_n := (P_n, T_n, A_n, AF_n, TP_n, TI_n, IP_n, OP_n, RP_n)$, where $n \in \mathbf{N}$, be the arbitrary PNTSs, $\forall i, 1 \leq i \leq n, \forall j, 1 \leq j \leq n: i \neq j \Rightarrow PTSN_i \angle PTSN_j$. The class $\boldsymbol{CPNET} \subset \boldsymbol{PNTS}$ then contains the following PNTSs:

i. if p be an arbitrary place, $p \notin (P_1 \cup P_2 \cup \ldots \cup P_n)$, then PNTS $BASE_p \in \boldsymbol{CPNET}$, where $BASE_p := (\{p\}, \emptyset, \emptyset, \emptyset, \emptyset, \emptyset, p, p, \emptyset\}$,

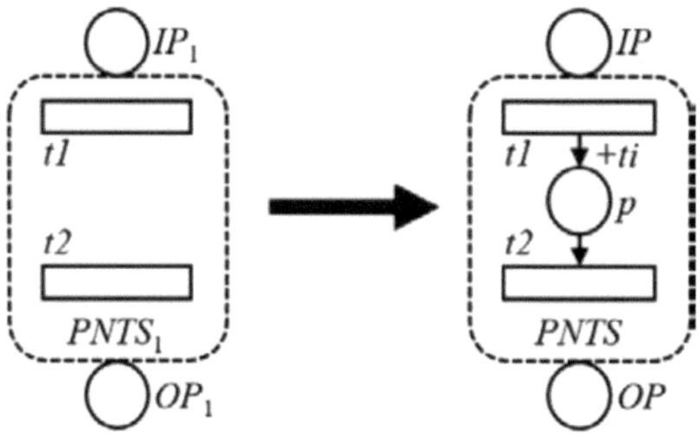

Figure 4. Symbolic representation of PNTS $PNTS := PNTS_1$.**JOIN**(p, $t1$, $t2$, ti).

ii. if $PNTS_1 \in$ ***CPNET***, $PNTS_2 \in$ ***CPNET***, t be an arbitrary transition, where $(t \notin T_1) \wedge (t \notin T_2)$, $ti \in \mathbf{N}_0$, then also $[PNTS_1, PNTS_2]$.**COMP**(t, ti) $\in$ ***CPNET***,

iii. if $PNTS_1, PNTS_2, \ldots, PNTS_n \in$ ***CPNET***, pi and po be the arbitrary places, $(pi \notin P_1 \cup P_2 \cup \ldots \cup P_n) \wedge (po \notin P_1 \cup P_2 \cup \ldots \cup P_n) \wedge (pi \neq po)$, ti and to be the arbitrary transitions, $(ti \notin T_1 \cup T_2 \cup \ldots \cup T_n) \wedge (to \notin T_1 \cup T_2 \cup \ldots \cup T_n) \wedge (ti \neq to)$, $ti1 \in \mathbf{N}_0$, $ti2 \in \mathbf{N}_0$, …, $tin \in \mathbf{N}_0$, $tio \in \mathbf{N}_0$, then also PNTS $PNTS \in$ ***CPNET***, where

$PNTS$:=$[PNTS_1, PNTS_2, \ldots, PNTS_n]$.**SYNC**$(pi, po, ti, to, 1, \ldots, 1, ti1, ti2, \ldots, tin, tio)$,

iv. if $PNTS_1 \in$ ***CPNET***, $p \notin P_1$ be an arbitrary place, $t1$ and $t2$ be the arbitrary transitions, $(t1 \neq t2) \wedge (t1 \in T_1) \wedge (t2 \in T_1)$, $ti \in \mathbf{N}_0$ and $af \in \mathbf{N}$, then also $PNTS \in$ ***CPNET***, where $PNTS := (P, T, A, AF, TP, TI, IP, OP, RP)$, such that:

- $PNTS := PNTS_1$.***JOIN***(p, $t1$, $t2$, ti),
- $\neg(\exists\, x_1 x_2 \ldots x_n \in \mathbf{CIRCUITS}_{PNTS}: (x_1 \in T) \wedge (x_n \in P) \wedge (n \in N))$. □

Four simple PNTSs $BASE_{P_1}$, $CPNET1$, $CPNET2$ and $CPNET3$ that are the members of the class ***CPNET*** can be seen in **Figure 5**, where:

- $CPNET1 := [[BASE_{P_2}, BASE_{P3}]$.**COMP**(**T2**, 2), $BASE_{P4}]$.**COMP**(**T3**, 5),
- $CPNET2 := [[BASE_{P6}, BASE_{P8}]$.**COMP**(**T6**, 2), $BASE_{P7}]$

  .**SYNC**(**P5**, **P9**, **T5**, **T7**, 1, 1, 1, 6, 3),
- $CPNET3 := [ANET2, ANET3, BASE_{P_1}]$.**SYNC**(**IP**, **OP**, **T1**, **T8**, 1, 1, 1, 4, 1, 8, 2)

  .**JOIN**(**P10**, **T3**, **T5**, 4).

  .**JOIN**(**P11**, **T3**, **T6**, 5).

Note also that the PNTS $CPNET3$ does not contain any circuit as it is required in the (4) of Definition 11.

Lemma 4. Let $PNTS \in$ ***CPNET*** be an arbitrary PNTS. Then $PNTS$ is pure-formed PNTS.

http://dx.doi.org/10.5772/intechopen.76769

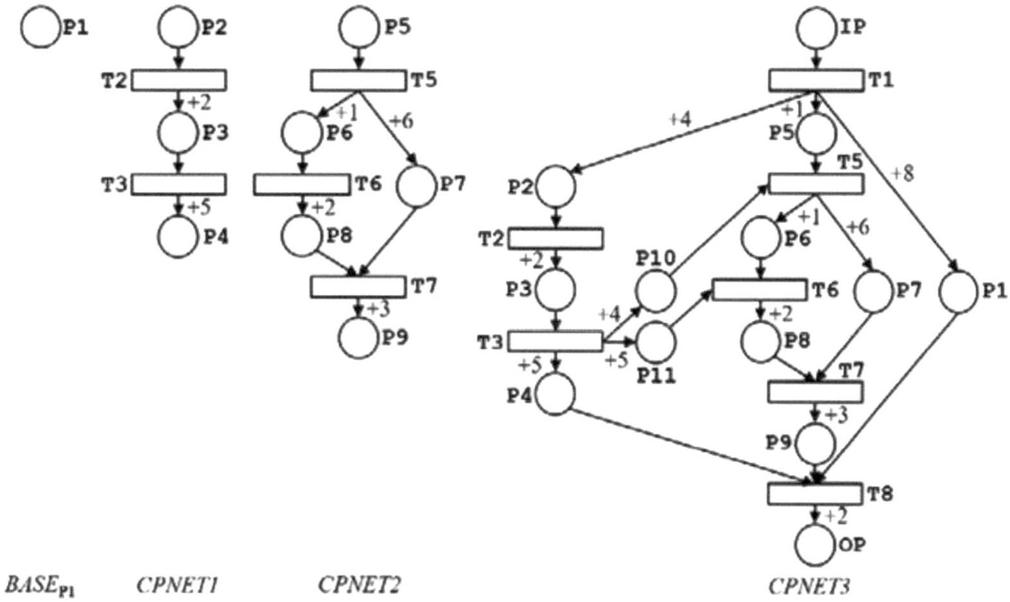

Figure 5. PNTSs $BASE_{P1}$, *CPNET*1, *CPNET*2 and *CPNET*3.

Proof. Clear, it directly follows from Definition 5, Definition 7, Definition 10 and Definition 11 and from the fact that any $PNTS \in$ ***CPNET*** does not contain any resource place (i.e., if the given *PNTS* is proper-formed PNTS, then it is also immediately well-formed and pure-formed PNTS). Furthermore, it is also clear that if we allow the existence of a circuit within the PNTS *PNTS* (see (iv) of Definition 11), there is always the danger of a deadlock in such a PNTS and *PNTS* is in this case neither a proper-formed PTSN. See, for instance, PNTS *CPNET*4 in its entry state S_e in **Figure 6**, where *CPNET*4 := *CPNET*3.**JOIN**(**P12**, **T7**, **T2**, 6). It is true that *CPNET*4 $\notin$ ***CPNET*** because there exists for instance the circuit.

$$\mathbf{T2\ P3\ T3\ P10\ T5\ P6\ T6\ P8\ T7\ P12} \in \boldsymbol{CIRCUITS}_{CPNET4}.$$

It is also clear that after the firing of the transition **T1** in the entry state S_e of the PNTS *CPNET*4 no one transition will be enabled for any value of the net time τ in this PNTS (i.e., there exists the deadlock marking in this PNTS) and thus the *CPNET*4 is not even proper-formed PNTS. □

Definition 12. The class ***CPPNET*** $\subset$ ***PPNTS*** contains all the PPNTSs *PPNTS* := (*PNTS*, S_e) where $PNTS \in$ ***CPNET*** and S_e := ((1, 0, …, 0), (<0>, <>, …, <>), 0). □

A_n example of a simple process is presented in the following paragraphs the characteristics of which will be studied with using of the PPNTS from the CPPNET class. The studied process is described in the following table of activities (see **Table 1**):

The CPM chart of the abovementioned process comprising the activities listed in **Table 1** is shown in **Figure 7** where:

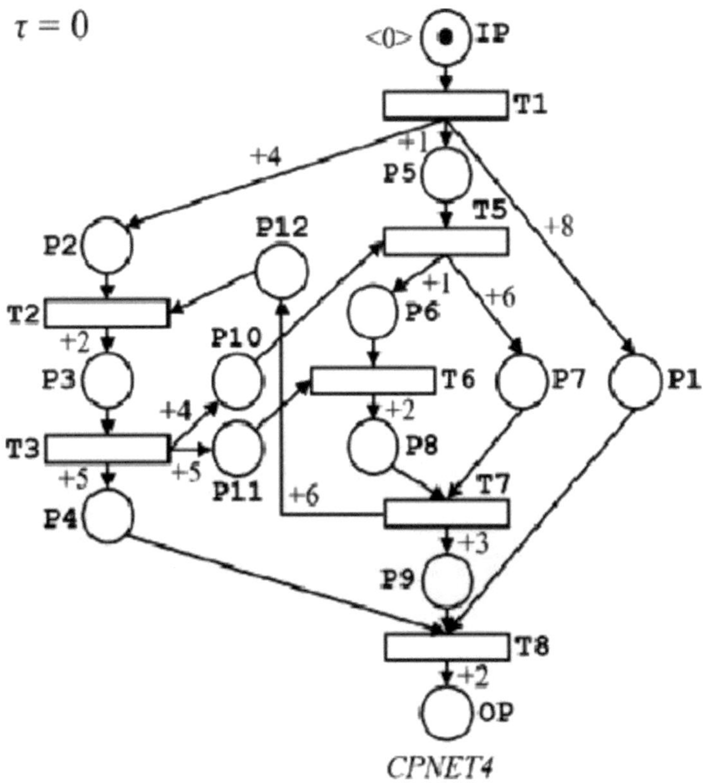

Figure 6. PNTS *CPNET4* in its entry state S_e.

Activity	Duration	Previous activities
A	2	–
B	3	–
C	4	–
D	2	C
E	6	B, D
F	5	A
G	5	C
H	3	E, F

Table 1. Table of activities and their dependencies of studied process.

- selected activity of the studied process and its duration is associated with each edge of the CPM chart,
- each node of the CPM chart is associated with its serial number (indicated in the upper half of the node), the earliest possible activation time of the given node (shown in the

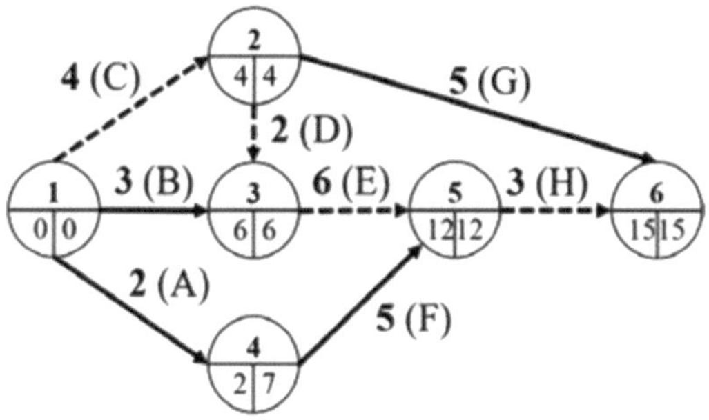

Figure 7. CPM chart of process activities listed in **Table 1**.

lower left quarter of the node) and the possible latest activation time of the given node (shown in the lower right quarter circle of the node),

- the desired links of the individual activities are expressed through the oriented edges and their associated nodes in the CPM chart,
- the critical path of the CPM chart passes through its nodes where the earliest possible activation time is equal to the latest possible activation time, that is, through nodes 1, 2, 3, 5 and 6, and it is thus formed by A, D, E and H activities (these activities are represented by dashed line graphs in the CPM chart) with a total duration of 15 time units.

The pure-formed PPNTS *PROC* that represents the process comprising the activities listed in **Table 1** can be seen in **Figure 8**, where

- $PROC := [PROC1, [BASE_{\mathbf{C}}, BASE_{\mathbf{G}}]$.**COMP**(**T5**, 5)].**SYNC**(**IP**, **OP**, **T1**, **T7**, 1, 1, 0, 4, 0),
- $PROC1 := [[BASE_{\mathbf{A}}, BASE_{\mathbf{F}}]$.**COMP**(**T3**, 5), $[BASE_{\mathbf{B}}, BASE_{\mathbf{E}}]$.**COMP**(**T4**, 6)]

.**SYNC**(**P1**, **H**, **T2**, **T6**, 1, 1, 2, 3, 3).

Places **A**, **B**, **C**, **D**, **E**, **F**, **G** and **H** of PNTS *PROC* represent individual activities of the studied process and the appropriate values of the time interval function *TI* then express the time durations of relevant activities (i.e., for instance, the time duration of the activity **A** is represented by the value of *TI*(**T2**, **A**) = 2, etc.).

In order to find the critical path of the process represented by PPNTS *PROC*, we first perform the association of each place and transition of the PPNTS *PROC* with the value of the critical path function *CP* that is introduced in the following Definition 13.

Definition 13. Let PPNTS $PPNTS := (P, T, A, AF, TP, TI, IP, OP, RP, S_e)$, $PPNTS \in$ **CPPNET**. The **critical path function** $CP: (P \cup T) \rightarrow \mathbf{N}_0$ is defined as follows:

i. $CP(IP) := 0$,

ii. $\forall p \in (P \setminus IP): CP(p) := CP(t) + TI(t, p)$, where $t = \bullet p$,

iii. $\forall t \in T: CP(t) := max(\{CP(p_1), CP(p_2), \ldots, CP(p_n)\})$, where $\bullet t = \{p_1, p_2, \ldots, p_n\}$, $n \in \mathbf{N}$. □

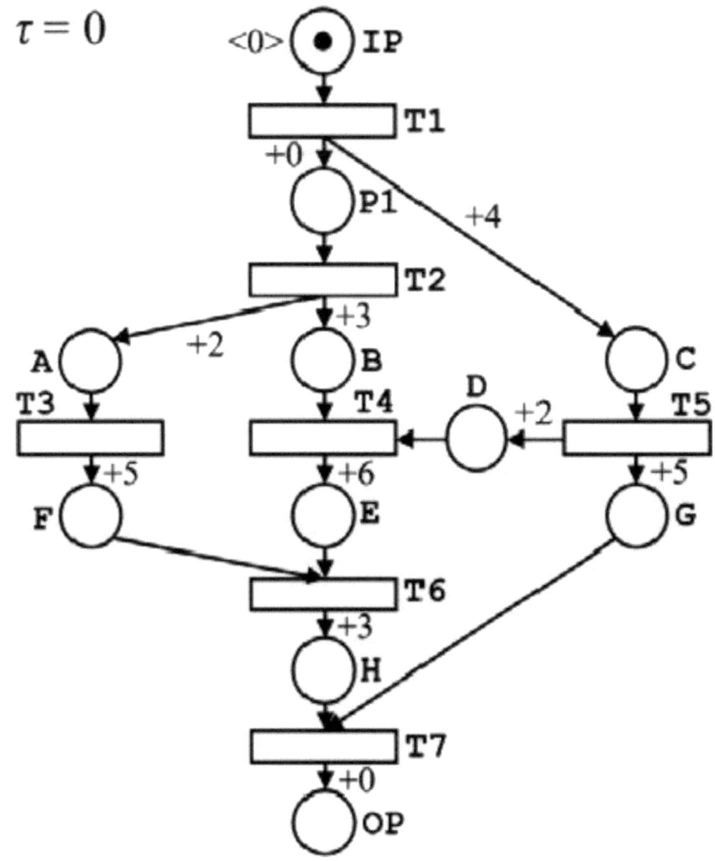

Figure 8. PPNTS *PROC* of process activities listed in **Table 1**.

The PPNTS *PROC* whose each place and transition is associated with the value of the critical path function *CP* can be seen in **Figure 9** (where, for instance, *CP*(**E**) = *CP*(**T4**) + *TI*(**T4, E**) = 6 + 6 = 12, where **T4** = •**E**; *CP*(**T4**) = *max*({*CP*(**B**), *CP*(**D**)}) = *max*({3, 6}) = 6, where •**T4** = {**B, D**}, etc.).

It follows directly from the Definition 4 that the value of the critical path function *CP* associated with any transition $t \in T$ of the arbitrary $PPNTS := (P, T, A, AF, TP, TI, IP, OP, RP, S_e)$, where $S_e := ((1, 0, \ldots, 0), (\langle 0\rangle, \langle\rangle, \ldots, \langle\rangle), 0)$, then represents the net time τ value when the given transition *t* will be fired. The value of the critical path function *CP* associated with the output place *OP* (i.e., *CP*(*OP*)) then immediately indicates the total duration of the process critical path (i.e., the net time τ value when the transition *t* = •*OP* will be fired). The algorithm for finding the set of PPNTS nodes of which the project critical path is formed is then obvious and it is expressed by the following pseudocode in PASCAL (the set of nodes forming the critical path of the project is then contained in the **CriticalPath** variable):

Node := *OP*; **CriticalPath** := {*OP*};

WHILE (**Node** <> *IP*) DO

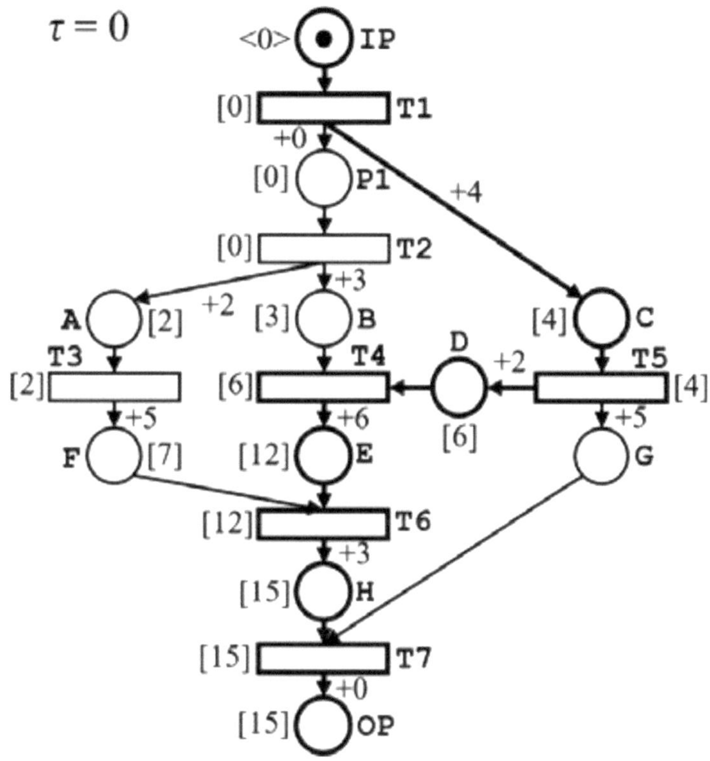

Figure 9. PPNTS *PROC* with the associated values of critical path function *CP*.

BEGIN

MaxValue := $max(\{CP(X_1), CP(X_2), \ldots, CP(X_n)\})$, where •**Node** = $\{X_1, X_2, \ldots, X_n\}$, $n \in N$;

Node := X_i, where $(X_i \in \{X_1, X_2, \ldots, X_n\}) \wedge (CP(X_i) =$ **MaxValue**);

CriticalPath := **CriticalPath** ∪ {**Node**};

END;

The critical path of PPNTS *PROC* is after applying of the above algorithm represented by the set **CriticalPath** := {**IP**, **T1**, **C**, **T5**, **D**, **T4**, **E**, **T6**, **H**, **T7**, **OP**} (see **Figure 9**). It is also clear that the given PPNTS may contain more critical paths with the same total time duration.

5. Conclusions

Further research in the field of PNTSs is mainly focused on the definition of additional unary, binary and *n*-ary PPPA operators preserving their specified properties, for instance, the binary **SUBST** operator that performs the substitution of the given PNTS for the selected place of another PNTS, and so on. In the field of the project management the research is focused on modeling complex processes, which individual activities can additionally share in parallel a selected set of the resources. These resources are then represented in the given PNTS by individual tokens located in the resource places of its selected net marking. Finding the time-optimal critical path of such a process as well as verifying the properties of the given PNTS that models such a process is generally a nontrivial problem and the use of PPPAs plays a crucial role here.

Another class currently being studied is the class of multiprocess Petri nets with time stamps that represents the generalization of the class of PNTS. The given multiprocess Petri net with time stamps then represents the finite set of processes each of which is modeled by a separate PNTS that share a common set of resources modeled by its individual resource places and their tokens. Many of the studied properties of the multiprocess Petri nets with time stamps are similar or identical to those of the PNTS class and they allow for a further generalization of the concept of a critical path formed by a sequence of activities of the process modeled by the given multiprocess Petri nets with time stamps.

Acknowledgements

This chapter was supported by the SGS at the Faculty of Economics, VŠB-TU Ostrava, under Grant Evaluation of comparative applications using cognitive analysis and method of data envelopment analysis, number SP2018/146.

Author details

Ivo Martiník

Address all correspondence to: ivo.martinik@vsb.cz

VŠB-Technical University of Ostrava, Ostrava, Czech Republic

References

[1] David R, Alla H. Discrete, Continuous and Hybrid Petri Nets. 2nd ed. Berlin: Springer-Verlag; 2010. 550 p. ISBN: 978-3642424694

[2] Reisig W. Elements of Distributed Algorithms: Modeling and Analysis with Petri Nets. 1st ed. Berlin: Springer-Verlag; 1998. 302 p. ISBN: 3-540-62752-9

[3] Diaz M. Petri Nets: Fundamental Models, Verification and Applications. 1st ed. London: John Willey & Sons; 2009. 656 p. ISBN: 978-0-470-39430-4

[4] Popova-Zeugmann L. Time and Petri Nets. 1st ed. Heidelberg: Springer-Verlag; 2013. 209 p. ISBN: 978-3662514351

[5] Furia CA, Mandrioli D, Morzenti A, Rossi M. Modeling Time in Computing. 1st ed. Heidelberg: Springer-Verlag; 2012. 424 p. ISBN: 978-3642323317

[6] van Hee K, Sidorova N. The right timing: Reflections on the modeling and analysis of time. In: Application and Theory of Petri Nets and Concurrency 2013, 34th International Conference Proceedings; 24–28 June 2013; Milan. Berlin: Springer-Verlag; 2013. pp. 1-20

[7] Martos-Salgado M, Rosa-Velardo F. Dynamic networks of timed Petri nets. In: Application and Theory of Petri Nets and Concurrency 2014, 35th International Conference Proceedings; 23–27 June 2014; Tunis. Berlin: Springer-Verlag; 2014. pp. 294-313

[8] van der Alst W, van Hee K. Workflow Management: Models, Methods and Systems. 1st ed. Massachusetts: MIT Press; 2004. 384 p. ISBN: 978-0262720465

[9] Huang H, Jiao L, Cheung T, Mak WM. Property-Preserving Petri Net Process Algebra in Software Engineering. 1st ed. Singapore: World Scientific Publishing; 2012. 318 p. ISBN: 978-981-4324-28-1

[10] O'Brien JJ, Plotnick FL. CPM in Construction Management. 8th ed. New York: McGraw-Hill; 2016. 736 p. ISBN: 978-1259587276